George Q. Huang · K.L. Mak

Internet Applications in Product Design and Manufacturing

W0037190

Springer-Verlag Berlin Heidelberg GmbH

George Q. Huang · K. L. Mak

Internet Applications in Product Design and Manufacturing

With 85 Figures

 Springer

Dr. George Q. Huang
Professor K. L. Mak
The University of Hong Kong
Department of Industrial
and Manufacturing Systems Engineering
Pokfulam Road
Hong Kong
P. R. China

e-mail: gqhuang@hkucc.hku.hk
e-mail: makkl@hkucc.hku.hk

ISBN 978-3-642-62819-1

Huang, George Q.:
Internet Applications in Product Design and Manufacturing / George Huang; K. L. Mak. – Berlin;
Heidelberg; New York; Barcelona; Hong Kong; London; Milan; Paris; Tokyo: Springer, 2003
(Engineering online library)
ISBN 978-3-642-62819-1   ISBN 978-3-642-55778-1 (eBook)
DOI 10.1007/978-3-642-55778-1

http://www.springer.de

© Springer-Verlag Berlin Heidelberg 2003
Originally published by Springer-Verlag Berlin Heidelberg New York in 2003
Softcover reprint of the hardcover 1st edition 2003

Typesetting: Camera-ready copy from authors
Cover-Design: Medio Technologies AG
Printed on acid-free paper   SPIN: 10875520   62/3020/kk   5 4 3 2 1 0

PREFACE

In recent years, the importance of Internet and World Wide Web (WWW) technologies in manufacturing industries has been rising very rapidly in a global context, the impact of which is deemed most profound ever since the Industrial Revolution. The waving interests in the electronic commerce and electronic business (e-commerce / e-business) have spread, from the heartland (product development) to the battlefield (shop floor), of manufacturing enterprises. The number of web applications is ever on the rise, and many practitioners are keen on trying these remote systems through web browsers to support their decision-making activities. Indeed, product design and manufacture professionals will soon be able to benefit from such remote services and supports commercially available on the Internet. The practice and performance of product development and realization are expected to make immense progress. Web applications in product design and manufacture signals the beginning of a new era of the digital manufacturing enterprise.

However, many loopholes are found in the development and application processes because of domain complexity and technology sophistication, thus generating new challenges to both the developers and practitioners. A simple example is the difference in the user interfaces between web applications and traditional applications. Indeed, abundant issues need to be resolved before the full launch of digital manufacturing can come into being.

Although numerous books have been published on electronic commerce / business, and on web / Internet technologies / programming, little has been gathered on developing and adopting web / Internet applications in product design and manufacture. These books either discuss the detailed programming skills and techniques or discuss generally about e-Commerce/ e-Business, without sufficient insights for web application developers and practitioners. This book therefore intends to fill the gap and by contributing literature significant at this stage of development in digital manufacturing.

This book is prepared to assist both developers and users of web applications to appreciate the potentials, as well as difficulties with suggested solutions, in developing and adopting web applications. Materials are selected from the results of various research projects that the authors have been conducting and supervising over the last five years. The objective is to equip potential users and practitioners of web applications involved in product design and manufacturing with a better picture of the technology, thus relieving bewilderment and enhancing confidence in application. In addition, web application developers and new researchers in this field are expected to gather better and clearer ideas regarding the selection of system architecture and design, development and implementation techniques, and deployment strategies.

The book is divided into two main parts and there are 14 chapters in total. The first part includes 7 chapters, discussing the overview and related issues of web and Internet applications in product design and manufacture. This introductory chapter has explained the concept of digital manufacturing enterprises. Chapter 2 reviews the recent progress in web applications and typical scenarios in product design and manufacture. Chapter 3 highlights some of the main challenges that have been learnt fro our researches in this field. Chapter 4 summarises a number of emerging issues surrounding the development and adoption of web applications for product design and manufacture. Chapter 5 focuses on the synchronization aspects of web applications. Chapter 6 is concerned with various business models of digital enterprises.

Part II includes 8 chapters. Each chapter presents one typical web application. The working scenario of each application is firstly described, and the system overview is then given, followed by the general application procedures. Special points are summarised toward the end of each chapter. Chapter 7 outlines a search engine dedicated to web applications in product design and manufacture. Chapter 8 discusses one of our first web applications for Design for Assembly analysis. Chapter 9 presents our efforts made and lessons learnt from web-based FMEA. Chapter 10 investigates how web application is used for managing engineering changes. Chapter 11 is mainly concerned with the development of the WeBid system for design chain management. Chapter 12 discusses collaborative product definition at the concept design stage. Chapter 13 presents our work on web-based design review and release management. Chapter 14 describes our development of the Teaching by Examples and Learning by Doing (TELD) courseware engine to support the teaching and learning in Product Engineering and Manufacturing Technology courses.

ACKNOWLEDGEMENTS

This book is based on the results from various research projects sponsored through a number of research grants awarded to the authors from various funding bodies, including the Hong Kong Research Grants Council, the Committee on Research and Conference Grants of the University of Hong Kong, Teaching Development Grants from Hong Kong Universities Grant Committee, Hong Kong/UK Joint Research Scheme, Hong Kong/Germany Joint Research Scheme, and the William Mong Engineering Foundation of the University of Hong Kong. These financial supports are gratefully acknowledged.

A number of research students, graduate research assistants, and post-doc research associates have worked on these projects. In particular, the authors are most grateful to Dr. Ming Nie, Miss Rachel Yee, Mr. Wilson Lee, Mr. Jun Shi, Dr. Jin Huang, Dr. Bing Shen, Dr. Xuebing Feng, Mr. Xianguo Chen, Mr. Yong Liu, Mr. Guanghui Zhou, Mr. Junying Shen, Mr. Jason Lau, Mr. Sheng Bin, Mr. Jianbing Zhao, Ms. Xiaolin Sun, and Dr. Zhuhua Jiang for their hard work in explorations into web applications in product design and manufacture. Their efforts make it possible for our ideas to be demonstrated on web sites, thus making it possible for the materialization of the present work. Supports from computer technicians in the Department of Industrial and Manufacturing Systems Engineering, the University of Hong Kong, are also greatly appreciated.

Finally, the authors wish to thank their families for their patience and dedicated support to help this book to turn into a reality.

ABOUT AUTHORS

Dr. George Huang is Associate Professor in the Department of Industrial and Manufacturing Systems Engineering, the University of Hong Kong. He obtained his B.Eng. degree in Mechanical Engineering from Southeast University, China, and Ph.D. degree in Mechanical Engineering from the University of Wales Cardiff, UK. He has been involved in teaching subjects related to product design and manufacture. His main research areas include Collaborative Product Commerce, Digital Manufacturing, focusing on the development and adoption of web and Internet applications in product design and manufacture. He has published extensively in these topics. Dr. Huang is a Chartered Engineer, and a member of IEE, ASME, IIE, and HKIE.

Professor K.L. Mak is Professor and Head of Department of Industrial and Manufacturing Systems Engineering, the University of Hong Kong. He obtained both his M.Sc.(Eng.) degree in Manufacturing Engineering and Ph.D. degree in Systems Engineering at the University of Salford in England. He is a Chartered Engineer, and had wide exposure to industry prior to joining the University of Hong Kong. He worked in some UK engineering enterprises, including the Pilkington Brothers Ltd., the T.S. Harrison and Sons Ltd., and some enterprises in Hong Kong. He has also built up strong linkages with local and overseas enterprises by acting as consultants for a number of industrial projects and system developments. At the University of Hong Kong, Professor Mak's current research interests focus mainly on Production and Operations Management, Product Development, and Manufacturing Systems Design and Control, and he has published extensively in these areas. He has also served the University in other capacities, for example, as Associate Dean of the Faculty of Engineering of the University of Hong Kong. He is also serving on the editorial board of a number of international journals.

LIST OF CONTENTS

1

INTRODUCTION

The Internet technology, with its capability of accommodating multimedia functions, has evolved rapidly during the last decade, and the World Wide Web (WWW or web) is the most popular and visible component among the multifarious devices in Internet technology. The client-server architecture also offers more channels for the sharing of information among project groups that are disparate in time and/or space. Indeed, the web has been widely employed in business, industry, the government, and academia.

A recent study indicates that the Internet technology has a significant bearing on the manufacturing industry to an extent comparable to that of the Industrial Revolution (Manufacturing Foresight 2020). The technology gradually transforms traditional manufacturing to what can be referred to as Digital Manufacturing (or e-Manufacturing), with more companies using the Internet to get into new markets, increase supply chain efficiency, create new value chains, and increase the efficiency of internal planning and operations.

Although the Internet technology has the potential of enhancing a company's competiveness, it has also created a new challenge for the manufacturing industry. Manufacturing companies have have just been persuaded to invest in Information Technology (IT) and Information Systems (IS), and have made appropriate changes to accommodate and adapt to these technologies. Now these information systems are becoming legacies even before their capabilities have been fully exploited. Now their roles have to be performed under the Internet and web environment. Hence, manufacturing companies have difficulties in formulating the most appropriate investment strategy in IT and IS under such rapidly changing environment.

In addition, the adoption of the web-based manufacturing approach is not only the decision of the manufacturers. If the customers, suppliers, and the commercial and financial sectors prefer to adopt the web approach, the IT industry will provide various web applications to meet their requirements. Eventually, most decision-support systems (DDS) for the manufacturing industry will also become web-based. They will then have to incorporate some or many of their business activities, decisions, and operations into web applications. Nevertheless, some of the questions that have to be addressed before any decision could be made are listed as follows:

- What types of product design and manufacture problems can the web approach handle most cost-effectively?
- What are the advantages of the web approach, as compared with the standalone and traditional client-server approaches?
- How should web applications be developed, implemented, deployed and applied to solve product design and manufacture problems?

This book therefore aims to provide stimulus to researchers, engineers and managers to the above questions, thus to assist individul organizations to reach the optimal decisions.

1.1.   WEB APPLICATIONS IN BUSINESSES

A web application is defined as any software application that depends on the World Wide Web, or simply web, for its correct execution (Gellersen and Gaedke, 1999). Hence, software systems that are explicitly designed for delivery over the web, for example web sites, and that use the web infrastructure for their execution, are web applications. For example, many information systems that were designed and built prior to the web are now wrapped and made available as web applications through the use of web browsers.

The web basically follows the client-server architecture. A web application normally includes two parts: the application client and server. However, it is necessary to distinguish web applications from general client-server applications. The main difference is the way in which application clients are installed and activated. In ordinary client-server applications, clients must be installed on specific client machines before they can be activated. Such installations are manually accomplished independent of web browsers. The case is dierrent for web applications. The client of a web application is automatically downloaded from the web server to the client machine. It is then installed and configured properly, and finally executed by the web browser at the client machine. In this case, the user may not necessarily be aware of the process because no installation or configuration is explicitly required. However, there are cases where certain amount of installation and configuration are stilled required from the web user so that the application client works properly. Nevertheless, these software systems are also loosely regarded as web applications.

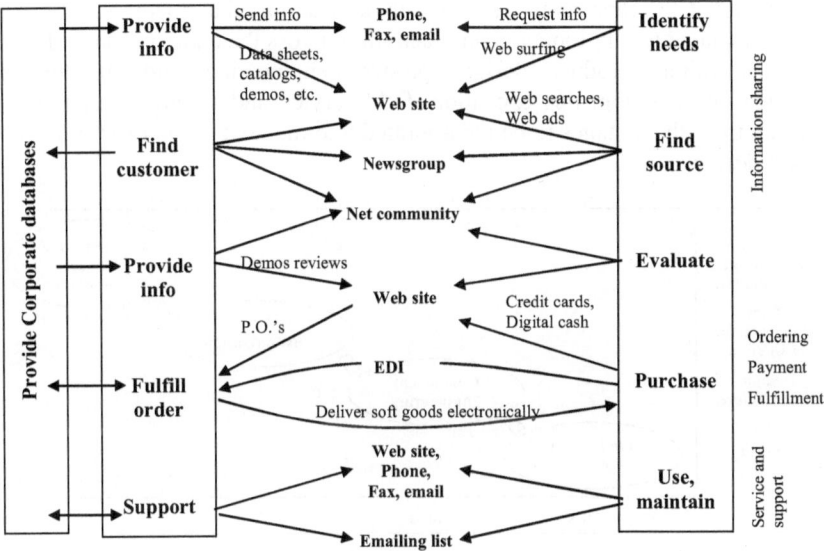

Figure 1.1 Typical business processes in e-Commerce supply chain lifecycle.
After Kosiur (1997)

Web applications have been increasingly used for the trading of products and services by handling the purchase and sale transactions and fund transfers over the Internet. This is commonly known as Electronic Commerce (e-Commerce). e-Commerce includes not only those transactions that centre on buying and selling goods and services to directly generate revenue, but also those transactions that support revenue generation, such as online marketing, order taking and processing, payment, support for delivery, after-sale support and facilitating collaboration between business partners. Figure 1.1 shows five typical buyer and seller processes in the e-Commerce life cycle.

The five business processes shown in Figure 1.1 are not, however, the focus of this book. Instead, this book concentrates on the roles that web applications play in the activities during the product development and realization process (simply Product Engineering). As shown in Figure 1.2, product engineering has shifted over the last 30 years from traditional product design to what is now termed as Collaborative Product Commerce (CPC) (Lee, 2000). During this period of evolution, suppliers and customers have become more involved in participating in the product development and realization process. The interfaces with these two external parties are the two ends of the supply chain where values are added most during the design stage.

Aberdeen (2000) defines CPC as "a suite of software and services that integrates several product-centric business processes across multiple independent enterprises into a single, closed-loop solution." CPC solutions are inherently web-based, and extensively use data sharing, collaboration and visualisation technologies. CPC represents a set of web applications that encompass business processes related to all product-centric activities across the entire product lifecycle, from the initial product design, product engineering and development, manufacturing and logistics execution, field service and technical support, and feedback from these stages to be incorporated into the next round of improvement product design.

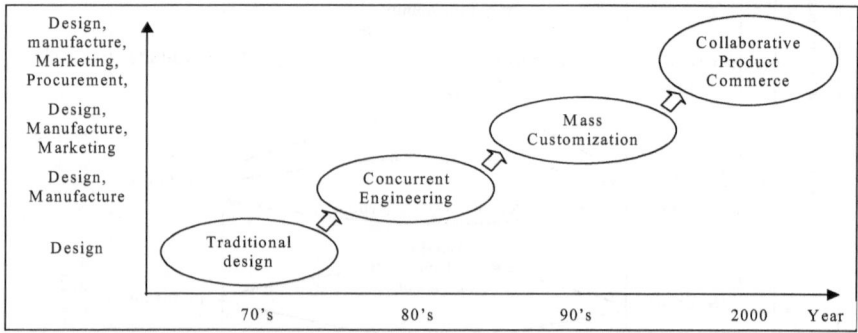

Figure 1.2 Participation of customers and suppliers in product engineering.
After Lee (2000)

CPC is complex, and the technology requirements are far more extensive than those associated with the e-Commerce shown in Figure 1.1. This can be understood from the following aspects:

- Some web applications in product development and realization are focusing on the design of new products (based on customer requirements), which are not available on the market yet.
- The transactions between players involve not only data and information, but also the exchange of knowledge.
- The transaction of information, both in terms of variety and intensity requires more complicated techniques. For example, 3D display and manipulation of geometrical information of products and processes on the Internet through the web remains a great challenge.
- The negotiation and collaboration between team participants within an enterprise (design, manufacturing, assembly, marketing, management, etc.) and/or across different enterprises (business partners, suppliers, customers, etc.) have a higher frequency and greater intensity when compared with ordinary e-commerce applications.

1.2.   ENTERPRISE PORTALS

In Chapter 2, it can be seen that a spectrum of web applications are available throughout the product development and realization process, focusing on topics ranging from market research to the supplies of tools/equipment and raw materials. If these web applications are left scattered on the Internet without a central hub, it will be very difficult for the designers and team members to locate the appropriate web applications at the right time for the right tasks. This situation is shown in Figure 1.3(a). The engineers should have an overall picture of the organization's business processes and functions. At the same time, they should also be conscious in the relevant documents, information flow and applications over the Internet. However, the situation will quickly become problematic when the number of information sources increases and the complexity of the business changes dynamically. The situation will be worse by the adoption of standalone and isolated applications, each with different working procedures, different interfaces which are non-self explanatory, and different installation and maintenance requirements.

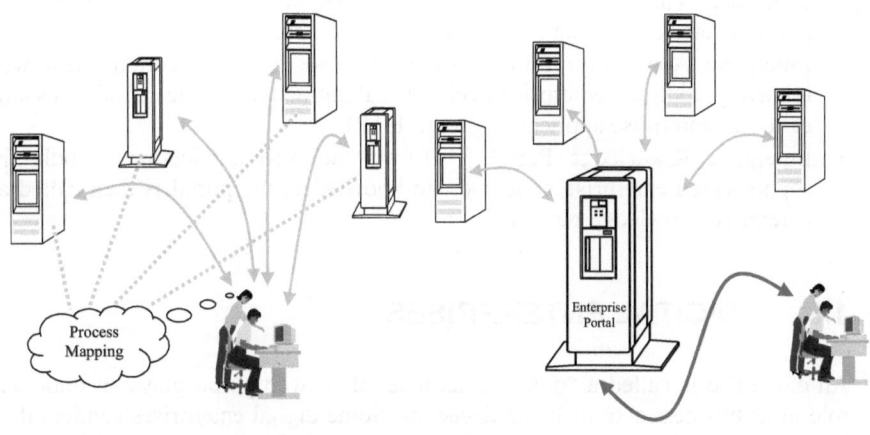

(a) Web applications without enterprise portal   (b) Web applications with enterprise portal

Figure 1.3 An enterprise portal as a central hub for accessing web applications.

The concept of enterprise portals is introduced as a central hub for accessing web applications to overcome the above problems, as shown in Figure 1.3(b). Enterprise portals gather and organise the huge amounts of unconnected data scattered across the enterprise, and then present the information in the forms of window areas, where each containing a view of the enterprise. Enterprise Portals provide a secure, personalized view of the enterprise for each individual user or class of users, based on job functions, roles or other relevant criteria. An Enterprise Portal integrates disparate sources of information and business logic, and provides a single point of access to the knowledge and processes of the organisation through a simple and intuitive interface to users of the extended enterprise, including customers, employees, suppliers, and business partners alike, with a single point of access to the knowledge and processes of the organization through a simple, intuitive interface. The technicalities of the back-end applications, documents or databases remain in the background and are invisible to users. Using an enterprise portal, these constituencies (customers, partners, employees) can conduct business via the Intranet, Extranet or Internet from wherever and whenever convenient for them.

The levels of complexity of the data that web applications deal with differ from application domains and problems. Some applications are only concerned with simple enterprise data. Some deal with relatively sophisticated enterprise information. Some may be involved in handling very complicated enterprise knowledge. Accordingly, there are three types of enterprise portals:

- Enterprise Information Portal (EIP): If these web applications are mainly concerned with enterprise data and information in simplistic contents, the portal is referred to as an enterprise information portal (EIP).
- Enterprise Application Portal (EAP): If these web applications deal with relatively complex enterprise decision-making activities, the portal is usually called an enterprise application portal (EAP).
- Enterprise Knowledge Portal (EKP): If the web applications handle the sophisticated enterprise expertise and knowledge, the portal is then called an enterprise knowledge portal (EKP).

1.3. DIGITAL ENTERPRISES

An enterprise is called a "digital enterprise" if its web portal plays an important role in its business activities and decisions. Some digital enterprises conduct their businesses mainly by providing online contents (which are relatively simple form of web applications) and more sophisticated web applications. Their business processes are entirely accomplished on the Internet.

However, traditional businesses such as manufacturing and even software development extensively employ web applications (and consequently digital enterprises) across the entire business processes to improve their operational efficiency and quality thus competitiveness. For example, a manufacturer becomes a digital manufacturing enterprise or simply an e-Manufacturing enterprise if it uses the web applications extensively in its business and engineering operations. Indeed, a digital neterprise has the following characteristics.

Firstly, the access to a digital enterprise is logically dimensionless in terms of time and geographical location, although it may be physically distributed (deployed) at different locations. To the user, it is a central web site. The geographical boundaries of the physical enterprise vanish with the emergence of the digital enterprise.

Secondly, a digital enterprise builds up the inter-connections between and among enterprises and intra-connections between and among units within a member enterprise via various levels of access control and authorities. The term "Virtual Enterprise" is often used to describe such a digital enterprise.

Thirdly, the lifecycle activities in the product development and realization processes are no longer arranged sequentially. The corresponding web applications are deployed in parallel across the entire the digital enterprise.

Finally, there is no longer difference between front- and back- offices in the digital enterprise.

A manufacturing enterprise can be divided into two parts: the physical enterprise (lower part of Figure 1.4) and the digital enterprise (upper part of Figure 1.4). The physical enterprise includes the shop-floor work centres that execute manufacturing operations to change the physical properties of the products, and the logistic transportation facilities that change the physical positions/locations of the products. Raw materials and semi-finished parts and components, which are inputs to the physical manufacturing enterprise, are transformed into finished products through a series of activities such as design, fabrication, assembly, inspection and testing, packaging, transportation, etc.

In order to optimise the performance of the physical enterprise, it is necessary to optimise firstly the plan of operation before it becomes operational and then its operation once it becomes operational. This optimised decision-making process involves digitally processing data, information and knowledge through corresponding web-based decision support systems (DSS). This environment is what we call "the digital enterprise". From the perspective of the digital enterprise, data are collected from various sources and then converted into meaningful (useful and useable) information to guide and optimise the decision-making activities in terms of market forecasting, product design, process planning, manufacturing resource planning, quality control, purchasing, etc. In contrast with the physical changes of product properties and positions in the physical enterprises, only logical changes take place in the digital enterprise in forms of data, information, knowledge, works, decisions, and of course their flows.

1.4.   MANUFACTURING PORTALS

A manufacturing portal, as shown in Figure 1.5, is a digital enterprise portal, which is used by one or shared among several manufacturing organizations or plants as a means to assist their decision-making activities. Physically, these plants can be dispersed at different locations. Organizationally, these plants can belong to the same corporation or to different corporations. In terms of information access and control, these plants appear at one location in the virtual space of the manufacturing portal.

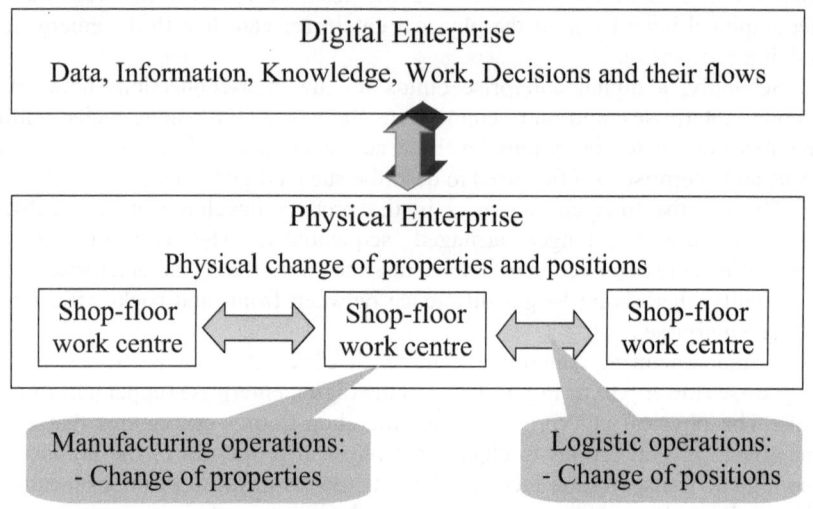

Figure 1.4 e-Manufacturing = Physical enterprise + Digital enterprise.

1.4.1.   Web Applications of Manufacturing Portal

A manufacturing portal is a central hub from which web applications are accessed for product design and manufacturing. These web applications vary from simple contents to sophisticated interactive decision-support systems. These applications will have a far more formidable effect when they ointly working together on the Internet through web browsers than working merely on their own separately.

The corporate portal of a manufacturing enterprise, as the name implies, encompasses all the web applications across the entire enterprise. It is an usual scene that those web applications related to product design and manufacture are grouped together to form sub-portals, corresponding to various functional departments. These sub-portals include:

- Marketing and Sales Portal,
- Procurement and Purchase Portal,
- Distribution and Logistics Portal,
- Human Resource Portal,
- Design and Development Portal,
- Production and Manufacturing Portal,
- Shop Floor Portal, and
- Finance Portal.

Each sub-portal takes the role of the corresponding functional department of the enterprise. In other words, they are departmental portals within the corporate portal. Normally, web applications within these departmental portals are mainly concerned with specific operations and activities of the departments. For example, web applications in the Design and Development Portal are those web-based systems that support decision-making activities in the product development and realization processes.

There are very few comprehensive manufacturing corporate portals currently available in the market. Most manufacturing portals are developed by extending one, or by merging several of the above sub-portals, resulting in several scenarios:

- CRM-oriented manufacturing portal
- SCM-oriented manufacturing portal
- PDM-oriented manufacturing portal
- ERP-oriented manufacturing portal

e-CRM (web-based Customer Relationship Management) minimizes the gap between a manufacturing company and its customers by improving its capability of engagement, transactions and fulfilment with its customers. The unique collection of applications available in e-CRM provides the company with open, adaptable, and user-friendly solutions that enable the company to become customer-centric and more efficient. These applications empower the staff by providing role-based workplaces, allow seamless real-time integration of front-office interaction and back-office fulfillment, and synchronize customer interactions across all channels. Hence, e-CRM helps enhance the company's sales and profitability, reduce churns, and delight the customers, employees, and shareholders.

e-SCM (web-based Supply Chain Management) facilitates a manufacturing company to to build fault-tolerant supply networks. This portal enables the company to anticipate and prevent the occurrence of shortages of items. Activated on the Internet, e-SCM extends supply chain technology to enable the company itself, its partners, and its customers to have instant access to critical business data such as orders, forecasts, production plans, inventory levels and key performance indicators.

ERP (Enterprise Resources Planning), evolved from the technology previously known as MRP (Materials Requirements Planning) and MRP II (Manufacturing Resources Planning), enables an organization to improve its operating efficiency by adapting to changing markets and customer needs. ERP is a set of software systems that provides key solutions to many problems in business operations and decisions. The applications cover a wide range: from market demand management through purchase and sales order management to production scheduling and shop-floor control. The challenge is to link these elements together seamlessly, and enable an organization and its partners to use the relevant information easily and proactively.

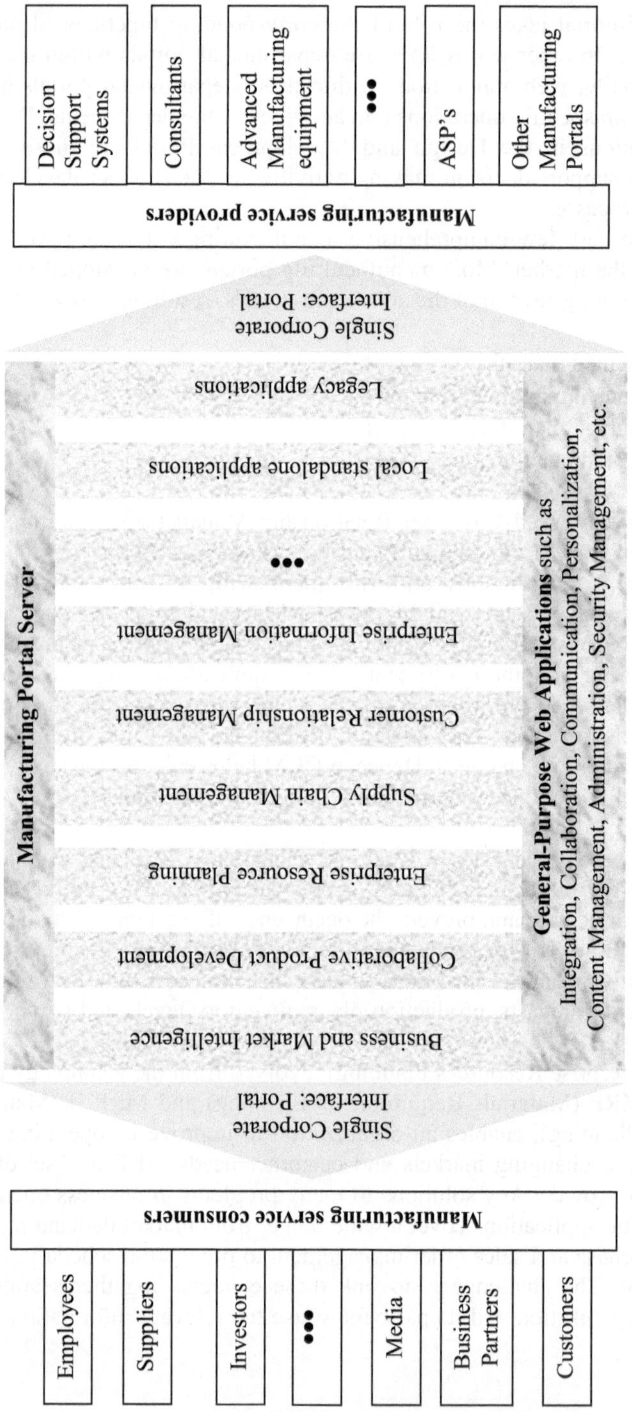

Manufacturing service providers

- Decision Support Systems
- Consultants
- Advanced Manufacturing equipment
- • • •
- ASP's
- Other Manufacturing Portals

Single Corporate Interface: Portal

Manufacturing Portal Server

- Legacy applications
- Local standalone applications
- • • •
- Enterprise Information Management
- Customer Relationship Management
- Supply Chain Management
- Enterprise Resource Planning
- Collaborative Product Development
- Business and Market Intelligence

General-Purpose Web Applications such as Integration, Collaboration, Communication, Personalization, Content Management, Administration, Security Management, etc.

Single Corporate Interface: Portal

Manufacturing service consumers

- Employees
- Suppliers
- Investors
- • • •
- Media
- Business Partners
- Customers

Figure 1.5 Manufacturing portal.

PDM (Product Data Management) is also widely known as Document/Drawing Management (DDM), Engineering Data Management (EDM), Enterprise Data Management (EDM), Product Information Management (PIM), and so on. No matter what acronym or term is used, it is an emerging technology embracing a suite of functionality for the management of product (including service and software products) definition information. Typical functions include project management (usually interfaces), product configuration management, workflow management, and document (hard copy, electronic copy or online form) management. Enterprise Knowledge Management is a high-level extension of PDM where data and information have been refined as indispensible knowledge over the period of time and accumulation of insights and skills.

1.4.2.   Manufacturing Portal Server

Among all the web applications in a manufacturing portal exists a special web application at its centre. This special web application is called the manufacturing portal server, which is simply a set of software solution on the Internet hosting web contents, services and applications. It is the integrator and controller of everything and everybody involved. It provides a Platform for Portal Operators to enable interaction between End Users and Application/Service Providers. The platform enables the procurement, and provision of Internet-based services.

The manufacturing portal server is implemented as a set of software components that can be executed on one or more server computers. A very small implementation can fit on one computer, but in most cases, a server farm will be deployed, to balance the load between multiple computers. The server is designed for maximum scalability and reliability, so that if a server box in a farm fails, its load is automatically assigned to another box.

At present, there are very few commercial manufacturing portal servers on the market, despite all the great potentials.

1.4.3.   Manufacturing Portal Users

Users of a manufacturing portal comprise different strata of parties or individuals who are participating in the product development and realization process. In most cases, a manufacturer categorises users as either internal and external. To its manufacturing portal, however, the difference is no longer between internal and external, but lies in the differing degrees of access authority. Such authorities are determined by their roles that they play during the product development and realization process.

No matter what privileges individual users are assigned with, they are divided into two groups:

• Information creators who not only need quick and easy access to the data, and the means to make substantial modifications to those data. Process planners, designers, analysts, and manufacturing engineers are some of the example users in this group.

- Information consumers who primarily need to view data and read/access related material. Example users include individuals in management, marketing, sales, support, suppliers, shop-floor personnel.

Information consumers need a low-cost, low-maintenance, and easy-to-use environment to view the information and perhaps add/publish simple attributes. For these users, the web-based server is the only viable solution.

The "information creators," on the other hand, will use the server as a means of communication and as a decision support system on high-end graphical user interfaces for concurrent design and collaborative engineering. This group of users demands a fast and versatile search and publishing capability, accessible from CAD and PDM systems. URLs are embedded in the databases that provide addition information for making modifications. Although a Web-based server may not be the only way to make data available to this group of users, the need for collaboration and information sharing at the extended enterprise level makes the Web-based solution very attractive.

1.4.4.   Application Providers and Portal Operator

The manufacturing portal operator is responsible for managing and maintaining the manufacturing portal. The interesting question here is: who should take the role as the portal operator? There are several options. For example, the manufacturing company itself can act as the portal operator. This requires the company to invest in the hardware and software, as well as the human resources. Alternatively, a few manufacturing companies form a group (consortium) to share the same portal, with the investment amount also shared out among the group members. Furthermore, the portals can be operated by specialist portal operators who not only have the hardware technologies and the skilled personnel can be invited to manage the portals.

Application Developers first develop the technology solutions or applications. The applications are then licensed or sold to the Application Providers, although applications developers may also serve as the application providers in some cases. The manufacturing portal server is a special web application and therefore the portal operator is a special application provider. The manufacturing portal operator subscribes to the services and applications on behalf of the users.

The manufacturing portal incorporates and hosts the applications as parts of its components in addition to those built-in web applications. Alternatively, the applications providers host the applications separately while links are incorporated in the manufacturing portal to provide access points, so that both the service providers and subscribers are able to access the services as the portal users.

Third-party applications providers are able to interact with a service aggregator, e.g. the manufacturing portal operator, after signing a contract with the operator and receiving a provider account and password.

1.5. BENEFITS OF DIGITAL ENTERPRISES

Potential benefits of Internet/web commerce and business have been widely perceived and documented in the literature. Most texts on e-commerce/e-business cover the benefits to some extent while some may dedicate entire chapters in this respect. For example, Chapter 2 of Timmers (2000) explores the potentials of B2B e-commerce in great detail. Therefore, a completely duplicated listing here is avoided. However, some aspects are worth brief mentions here again in the context of digital manufacturing.

Firstly, with the client-server architecture, both web-based design services and their users can be geographically distributed anywhere in the world as long as they are available on the Internet. This suits well with collaborative product development where team members often work at different localities and on different shifts.

Secondly, web applications are accessible openly and concurrently 24 hours a day throughout the world. Such open accessibility reflects the ready availability of specialist skills and knowledge required in collaborative product development.

Thirdly, as long as the user has the use of an open standard web browser in a client on the Internet/Intranet, he or she can have instant access to any web-based design tools. Both the client and the server communicate to each other using a standard HTTP (Hyper Text Transfer Protocol), regardless of their hardware configurations and operating systems.

Fourthly, installation, maintenance and upgrading are no longer necessary on the client side. These activities are accomplished on the server side by the service providers. Installation is automatically achieved during the downloading process when an access is made to a web site.

Fifthly, web applications have the same performance as standalone systems in terms of functionality, interactivity and usability. This is owing to the multimedia capability and client side scripting/processing of the web technology.

Sixthly, web applications can perform faster than conventionally networked servers because some computation is performed locally on the client machines rather than remote machines.

Seventhly, unlike standalone systems where only single users can gain access at a time, web applications can be accessed by multiple users at the same time. This truly creates a concurrent engineering environment where product development activities can be carried out in parallel.

Eighthly, web applications possess greater scalability. This can be easily understood using the 3-tiered architecture. Web applications can either share the same data source or have their own data sources at the database tier. The server components of the web applications can be freely deployed without affecting each other. The client components (web pages) can be arranged (scaled up) as desired.

Finally, when the web is used for information management, changes can be posted on the network, thus allowing users in remote locations to have instant access to these changes. A dynamically generated web page that reports any relevant information to the manufacturing engineers, either on request or by notification, could drastically reduce the "search" time. Furthermore, there is no need for the user to know explicitly how the data is transferred in the system.

2

RECENT DEVELOPMENTS OF WEB APPLICATIONS IN PRODUCT DESIGN AND MANUFACTURE

Product design and manufacture has been traditionally an area for intensive research and extensive application of computer systems. Early developments in Computer Numerically Controlled (CNC) and programmable machinery and devices have brought significant improvements in productivity and product quality. An issue of "Islands of Automation" had emerged before further potentials could be achieved. By the late 70s, the need for integration was recognised and one of the major challenges faced by the CIM (Computer Integrated Manufacturing) research was to address the conspicuous lack of standards for communication capability among computers and programmable devices. Later on, MAP (Manufacturing Automation Protocols)/TOP (Technical and Office Protocols) were proposed and gradually developed as international standards (Jones, 1987).

In the meanwhile, considerable advances have been achieved in the development and application of a variety of computerised decision support systems in product design and manufacture. For example, Computer Aided Design (CAD), Computer Aided Manufacturing (CAM), Computer Aided Process Planning (CAPP), Computer Aided Shop Floor Control (CASFC), and Computer Aided Production Management (CAPM) systems are evolved through several generations. In particular, they have recently been enhanced by intelligent and knowledge- based expert system technology. As a result, their usability, functionality and performance have been greatly improved. However, a new issue of "Islands of Expertise" has emerged. The exchange of information and knowledge between these systems becomes a bottleneck. This is a new challenge for CIM research at present.

The nature of "Islands of Expertise" is characterised by the geographical distribution over time of project team members who are equipped with computerised decision support systems. Significant efforts have been made to develop computer supports in order to facilitate teamwork. A snapshot of the early developments was taken in a special issue of the IEEE Computer Journal (Computer, 1993). More recently, several major initiatives and projects on collaborative product development in America and Europe have adopted the Internet/intranet technology as their collaborative engineering infrastructure. The World Wide Web (web or WWW) technology, in particular, offers a number of promising features. The client-server web architecture offers tremendous opportunities for sharing information among product development team members who may well be distributed in terms of both time and space. The multimedia capability of the web can be used to develop web-based applications that are equivalent to standalone systems in terms of functionality, performance and usability.

A web application can be defined as any software application that needs the World Wide Web for its correct execution. A variety of web applications have already been developed in product design and manufacture, and more are yet to appear in the very near future. The web technology is highly expected to provide an adequate information infrastructure to support the collaboration between work-centres geographically distributed over time.

This chapter will examine a number of major initiatives and projects recently completed or just launched on web-based product design and manufacture. Typical application scenarios as reported in the literature will be discussed. It is hoped that this chapter will be able to assist the researchers who have recently entered, or have worked for some time on this field of web application in product design and manufacture to better understand the potentials and challenges of the web technology.

2.1.   WEB APPLICATIONS IN PRODUCT DESIGN AND MANUFACTURE

One of the first and most significant initiatives in the development and application of web-based systems in product design and manufacture is the substantial American research project – the MADE (Manufacturing Automation and Design Engineering) program. MADE is a DARPA (Defence Advanced Research Projects Agency) program initiated in 1992 and subsequently completed in 1996. This program involved a number of major research centres or groups, resulting in valuable publications at conferences, journals, and on the Internet (Cutskosky et al, 1996; Petrie, 1996; Whitney et al, 1995; Bryant et al, 1995; Will, 1996). The MADE program supports research, development and demonstration of enabling technologies, tools, and infrastructure for the next generation of design environments for complex electro-mechanical systems. It concerns with the comprehensive information modelling and the design tools needed to support rapid design of electro-mechanical systems. It emphasises the notion of "tag team" design, in which each designer performs the functions he or she is most expert at while leaving behind enough information in design information web for other designers to pick up. MADEFAST was a demonstration of this approach conducted by several research groups who collaborated in the design and manufacture of a prototype sensor array aiming system. The MADE program continues as RaDEO (Rapid Design Exploration and Optimisation) program (RaDEO, 1997).

Since then, there have been rapid developments of web applications in collaborative product development. This section categorises recent developments under four directions:

- Individual web-based decision support systems which are more conducive to product design and manufacture.
- Individual web-based decision support systems that interact with each other.
- Web applications that are especially designed and developed to facilitate and support group or team work in collaborative product development.
- Enterprise portal platforms that host and integrate various web applications into a central corporate hub.

2.1.1. Individual Web-Based Decision Support Systems

Web technology enables us to develop virtually all types of decision support systems in product design and manufacture. Owing to the multimedia capability of the web, web-based systems are functionally at least equivalent to, if not better than, those standalone counterparts. Client side scripting renders web-based systems to perform highly interactively and execute simple computation quickly. Now, the question is not whether web applications are as good as standalone systems, but how can the existing standalone legacy systems be converted into the web-based systems so that they can be used on the Internet. This area obviously deserves further investigation, and it will be mentioned again in the next section. A number of significant developments and major projects will be selected for explanation below.

Starting from the front-end of the product development process, the web-based approach is particularly suitable for customer requirement management or market research. An interesting work at the Philips Advanced Development Centre was to use the Internet as a communication infrastructure for lead user involvement in the new product development process (Muller et al, 1996). Another related project was to use the web for customer requirement analysis for software product development (Anton and Liang, 1996).

The web technology is playing an increasingly important role in marketing and sales, and after-sale customer services. Web-based product and component cataloguing is an example. Huang and Mak (2001c) reported on a prototype web-based product cataloguing system. Active Catalogue (Will, 1996) is another project aiming to develop web-based component catalogues including models to enable "try before you buy" simulation analysis during the product development process. HKCAINS (Hong Kong Accessory Information Network System) is an industry-based project, aiming to develop a prototype network for Hong Kong apparel industry, and which in particular, for the retrieval of accessory-related information and seek advice from the apparel industry on the economic viability of establishing such a network (HKTAIGA, 1996). Wong, Veeramani and Wong et al (1996) proposed methodologies for rapid and accurate response to request-for-quotation and they demonstrated the method on the Internet. Web sites are also ideal for setting up virtual customer service centres. One potentially significant project is "Intelligent Product Manuals" (Pham, 1998). The objective of this project is to supply the necessary electronic information to support the continued use and maintenance of a product from its delivery to its disposal.

In design and manufacture, Wagner et al (1995) experimented with developing fixture design systems on the Internet. One of the early web applications was to provide rapid prototyping services on the Internet (Bailey, 1995; Wright and Burns, 1997). Smith and Wright (1996) collected a number of web-based design and manufacturing services for their CyberCut experiment. Roy et al (1997) have presented experimental workbenches for web-based design to production, including activities such as conceptual and detail product design, Process planning, Design for Manufacture, NC programming and rapid prototyping. Kim et al (1998) proposed a web-based architecture for collaborative design in mixed platforms and dispersed geographical environments. It used open data standards to allow users on a wide variety of platforms to access and visualise product information. Kalyanapasupathy et al (1998) proposed to use the Internet to support the generation of Group Technology codes for mechanical parts. IAMS (Intelligent Assembly Modelling and Simulation) aimed to facilitate assemblability checking in a virtual, simulated environment in order to avoid expensive and time-consuming physical mock-ups. It is, in fact, a project within a broader effort of developing Collaborative Open Design System (IAMS, 1998).

In process and production planning and control, Yen explored the web-based simulation (Yen, 1997a) and production scheduling (Yen, 1997b). The IPPI (Integrated Product Processing Initiative) project is another major effort aiming to develop and validate a prototype process planning system that will utilise form feature product models defined in STEP format and be capable of generating intermediate product models which represents the state of the product prior to and subsequent to each manufacturing operation. The goal of the IP3S (Integrated Process Planning/Production Scheduling) project is to dynamically convert standards-based product specifications into process plans and schedules that best accommodate the current shop load, the status and allocation of machines, fixtures and tools, and raw material availability, while minimising production costs, lead-times and inventories and maximising due date performance simultaneously. These projects make extensive references to STEP and CORBA as their ultimate operational constructs.

From a wider perspective of supply chain management, the web technology also has a great potential. Early applications have been focused on implementing sophisticated mathematical decision models for supplier selection. OSPAM (Optimal Selection of Partners in Agile Manufacturing) was probably one of the first attempts to apply extensively IT/IS in general and the Internet in particular in supplier selection (Minis, 1995). Vanwelkenhuysen (1998) has described a Tender Support Expert System for industrial centrifugal pumps. The system assists sales engineers to quickly generate and explore valid pump configurations technically as a response to customer requirements. Kroemker et al (1997) presented a concept of simultaneous bid preparation and implemented a prototype infrastructure to support interdisciplinary co-operative bid preparation over a distributed heterogeneous system environment. These were only the beginning and more work remains to be done in this direction.

Huang and Mak (2000b) reported a research project with the aim to develop an overall methodology for enabling better supplier involvement in the new product development process and to demonstrate the framework through a prototype web-based platform on the Internet/intranets using the web technology. Four focus areas have been identified: (1) To develop a product-oriented supply chain model, which is consistent with the new product development process; (2) To develop a mechanism for the customers to invite and potential suppliers to submit bids for manufacturing specific product components respectively; (3) To develop a rigorous but pragmatic supplier selection methodology; and (4) To develop a mechanism for facilitating information and task sharing between the customers and suppliers.

2.1.2. Inter-Operable Web Applications

Web applications are usually developed for human users to use with web browsers. Some web applications have been developed in such a way that they can access each other with little or without human intervention. This type of web applications is considered as inter-operable. For example, an application server and a client are inter-operable. The client is able to initiate or terminate the server. More importantly, the server and the client are developed in the way such that they can exchange information when they are operating in a mutually understandable way.

However, such mutual understanding is not guaranteed between individual web applications. The reason is that participating web applications in a project are usually developed by third-parties or by the same developer but at different times. At the time when a web application is developed, other web applications may not exist or are still unknown for their existences. Therefore, they are not developed to provide "plug and play" type of mutual inter-operation.

Considerable efforts have been made to provide a standard for developing inter-operable Internet applications. Three of the most popular distributed object paradigms are Microsoft's Distributed Component Object Model (DCOM), OMG's Common Object Request Broker Architecture (CORBA) and JavaSoft's Java/Remote Method Invocation (Java/RMI). The standards for distributed computing only provide specifications, regarding the computational feasibility in terms of the format of information and control exchange between applications. They do not deal with the technical contents of the exchange. Both the formats and contents are of great significance. In the area of Distributed Artificial Intelligence, KIF (Knowledge Interchange Format) has emerged as a format for KQML (Knowledge Query and Manipulation Language) arguments. In the area of product design and manufacture, STEP (STandard for the Exchange of Product model data) is being developed and adopted by individual participants as their internal formats for representing product, process and resource data, and also their results.

As far as the contents of exchange messages are concerned, there has been effort in developing engineering ontologies. This effort focuses on defining formal vocabularies for representing knowledge about engineering artefacts and processes. These vocabularies specify the assumptions, underlying the common views of such knowledge.

However, majority of existing web applications in the field of product design and manufacture are not developed as being inter-operable. This has been highlighted by early experiments such as CyberCut and MADEFAST. CyberCut is an extension of the Integrated Manufacturing And Design Environment (IMADE) developed at the University of California-Berkeley into a distributed agent environment on the Internet (Smith and Wright, 1996). Another illustrative example system is MADEFAST. It was an early example of a new and rapidly growing genre of projects that use the World Wide Web (WWW) extensively for collaborating and achieving results. The basic idea behind the MADEFAST project is that an engineer would have access to a powerful workstation for recording designs, sketches, memos, and meeting notes, etc. This workstation is also connected to the Internet, where it has access to the shared MADEFAST project pages posted by all participants, as well as tools and services.

Most participant systems included in these CyberCUT and MADEFAST experiments were developed by third-parties or by the same developer at different times. Therefore, they are not interoperable. Further processing is necessary. One solution is to introduce the concept of agents that wraps up web application, even standalone applications, so that they can be inter-operable. Agents are usually attached to the corresponding web applications on the server-side but downloaded to the client-side. Such downloaded agents connect the clients to the corresponding web applications. Frost and Cutkosky (1996) and Smith and Wright (1996) explained how individual agents work and how they work as a community. The authors are further extending the concept of intelligent agents in the context of workflow management (Huang, Huang and Mak, 2000a).

2.1.3.   Web Applications for Group/Team Work

Significant progress has been achieved in developing and applying support systems for group or team decision-making. There have appeared two major research themes. One is generally referred as Computer Supported Collaborative Work (CSCW) and the other is Workflow Management. The web technology has been used in both.

The aim of the web-based CSCW research is to develop web-based framework or architecture to support teamwork or group decision-making, rather than individual decision support systems (DSS) for solving particular problems. The participants in these frameworks are usually human members of a project team. Much of the decision- making is accomplished by the individual participants, with or without the help of computerised DSS. One example of web-based CSCW is GroupSystems Web (Romano et al, 1998). It is an HTML/JavaScript web-based Group Support System. It provides an environment for group coordination and a suite of collaborative tools. The environment builds upon the GroupSystem concept, which provides a computer for each participant, software for each task, a public screen to focus attention, a network to share information, access to external data at anytime, at any place supports and extends that concept to provide support for distributed collaboration.

The research on workflow management seems to involve not only human participants, but also software systems. Systems are able to initiate and terminate by themselves. In contrast, participants in CSCW are human users who may be assisted by computer systems, not the software participants. In a workflow model, participants, whether humans or software, are represented as nodes and the flow of work as edges. The flow of work includes the flow of data and the flow of control. WebWork (Miller et al, 1997) is an example of web-based workflow management system. It provides the command, communication, and control for the individual tasks in the workflow. WebWork implementation relies solely on the web technology as the infrastructure for the enactment system. It supports a distributed implementation with multiple web servers. It has been developed as a complement to its more heavyweight CORBA-based counterparts with the goal of providing ease of workflow application development, installation, use, and maintenance.

Although software systems can be participants in the web-based collaborative workflow management systems, they can be operated manually by human users or automatically operated by other systems. In the latter case, the software participants become inter-operable agents as discussed in Section 2.2. The authors have proposed an approach where participant systems are represented as intelligent agents and their inter-related activities are controlled and scheduled by flows of the work (Huang, Huang and Mak, 2000a).

2.1.4.   Enterprise "Portalets" of Web Applications

Enterprise portals gather and organise the huge amounts of unconnected data scattered across the enterprise, and then present the information in the forms of window areas, where each containing a view of the enterprise. Enterprise Portals provide a secure, personalized window into the enterprise for each individual user or class of users, based on job functions, roles or other relevant criteria. Enterprise Portals integrate disparate sources of information and business logic, and provide the user of the extended enterprise, including customers, employees, suppliers, and strategic partners alike, with a single point of access to the knowledge and processes of the organization through a simple, intuitive interface. The technicalities of the back-end applications, documents or databases remain in the background and are invisible to the user. Using Enterprise Portals, these constituencies (customers, partners, employees) can conduct business via the Intranet, Extranet or Internet from wherever they are and at any time that is convenient for them.

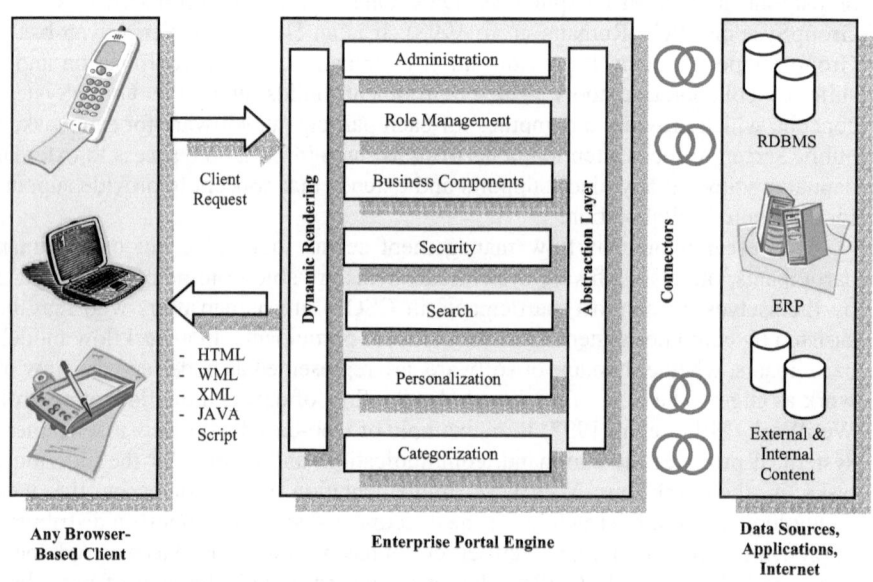

Figure 2.1 Architecture of the **appsolut** Enterprise Portal Suite.

The so-called portal products range from the simple Intranet indexing and search tools to the enterprise-level database-driven information storage and retrieval products. Figure 2.1 shows the architecture of the appsolut Enterprise Portal Suite (http://www.appsolut.com).

An enterprise portal consists of a set of enterprise web applications that are interconnected according to some logics. If these web applications are mainly concerned with the enterprise data and information in simplistic contents, the portal is referred to as an enterprise information portal (EIP). If these web applications deal with relatively complex enterprise decision-making activities, the portal is usually called an enterprise application portal (EAP). If the web applications are about the sophisticated enterprise expertise and knowledge, the portal is then called an enterprise knowledge portal (EKP). Enterprise Information Portal (EIP) must be capable of content processing, filtering aggregation, management and personalisation, and offer basic repository service. It should also contain the search and retrieval function and present the information in categories and classes. Enterprise Application Portal (EAP) is all about integration, aggregation and presentation of selected ERP (enterprise resource planning) capabilities. Enterprise Knowledge Portal (EKP) helps the users to decide the fate of their business by providing the updated business information with or without some past knowledge of the enterprise.

Most of the enterprise portalets are for building enterprise portals from general-purpose web applications. They are not specifically designed for building up engineering portals from design and manufacturing web applications. The need for this, however, has been recognized and efforts are gradually being made in providing special-purpose enterprise portalets for product design and manufacture. For example, CAD (Computer Aided Design) and PDM (Product Data Management) vendors become major players of CPC web applications and enterprise portalets. For example, Windchill from PTC (Parametric Technology Corporation) is one of the leading comprehensive CPC portalets (http://www.ptc.com/products/windchill/). Furthermore, Windchill also supplies a rich set of CPC web applications or FACTORS, ranging from a general-purpose search engine, collaboration platform, to manufacturing planning and supply sourcing.

2.2. TYPICAL SCENARIOS OF COLLABORATIVE PRODUCT COMMERCE

As mentioned in Section 2.1, web applications have appeared in almost every stage and area in the new product development. Some are in their infancy stages, some are their concept testing stages, and some are of professional standards on the market. This section briefly introduces several typical scenarios that the researchers and practitioners have been working on over the last few years. It is not intended to be an exhaustive list, rather to add some motivating flavour.

Figure 2.2 Web application for market testing of product concepts.
Source: http://www.surveysite.com/survey/conjoint/conjoint-example.html

2.2.1.   Market Research and Concept Testing

Web applications make it possible to carry out online market research and product concept testing. Figure 2.2 shows an example of carrying out a Conjoint Analysis at a web site (http://www.surveysite.com/survey/conjoint/conjoint-example.html). Dahan and Srinivasan (2000) developed an Internet-based product concept testing method, which incorporates virtual prototypes of new product concepts, and substituting them for physical prototypes. The method can be used with either static representations of the products or dynamic representations that demonstrate how the product works through a simulated video clip. The objective of this method is to allow design teams to select the best of the new concepts within a product category, with which to proceed then there is no need to develop physical prototypes.

The general procedure is as follows:

- The manufacturer sets up a web site for customers to voice their requirements.
- The manufacturer (product development team) reviews the customer requirements in order to establish design specifications for the new product.
- The design specifications (customer requirements) are then used to formulate design concepts (this is conceptual design to be discussed in the next section).
- Virtual (and/or physical) prototypes are prepared for candidate concepts and displayed on the Internet.
- Customers are invited to test these concepts at the web site.
- The team on the manufacturer's side reviews the customer responses and proposes changes to the conceptual design (to be discussed in a subsequent section).
- The project proceeds to the next stage.

As virtual prototypes cost considerably less to build and test than their physical counterparts, design teams using Internet-based product concept research may be able to afford to explore much larger number of concepts. Virtual prototypes and the testing methods associated with them may help to reduce the uncertainty and cost of new product introductions by allowing more ideas to be concept- tested in parallel with target consumers.

2.2.2.   Collaborative Early Product Definition

Early product definition, also known as conceptual product design and product conceptualization, is a collaborative effort of the team members. Web applications are particularly attractive. While Chapter 12 is dedicated to this topic, the SchemeBuilder is an example of computer aided tool for product conceptulization, as shown in Figure 2.3. Although it is not yet a web-based tool, it facilitates the development, refinement and selection of design concepts through a collaborative effort.

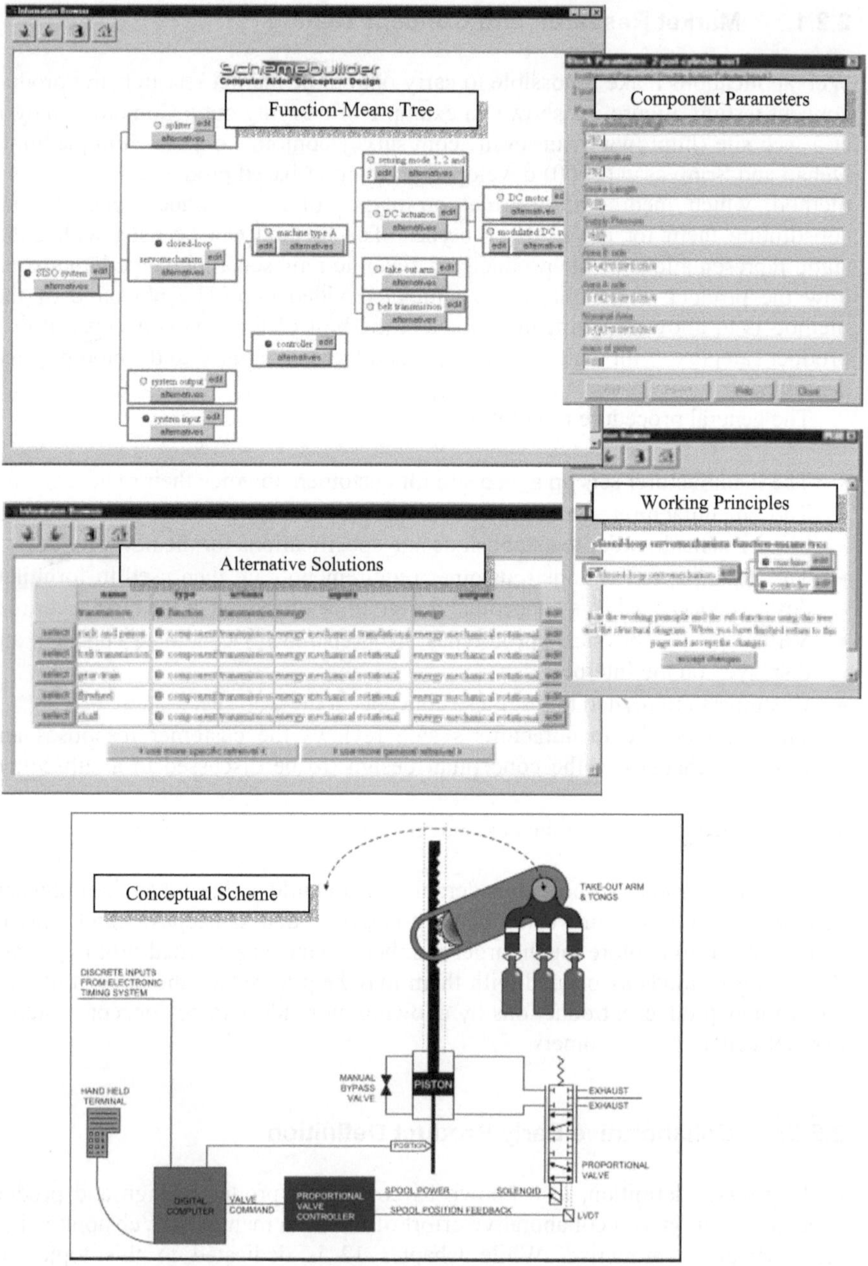

Figure 2.3 SchemeBuilder for conceptual product development.
Source: http://www.comp.lancs.ac.uk/edc/schemebuilder/

Schemebuilder is a software tool which enables the rapid development of conceptual product designs, known as schemes. The computer helps the user to explore alternative concepts and produce design simulations. The tool provides a design synthesis environment which is coupled with a structured Knowledge Base. The Knowledge Base provides intelligent access to design knowledge and a component database. This is integrated with Simulation Environment for design analysis, capable of cross-domain, object-oriented simulation.

Both abstract and concrete knowledge are represented. The knowledge and experience of any user may be added. Case Based Retrieval is used with multi-dimensional, hierarchical indexing. Three types of knowledge are represented: Means of achieving functions, Working Principles or given function structures which experience has shown to be successful, and Components which embody the means. Very complicated knowledge may be represented as rules. Advice, triggered when relevant circumstances arise, may be accepted or ignored. Control advice has been embedded which uses the automatic simulation capability. Schemebuilder is capable of automatic simulation generation made possible by the use of object-oriented models. The simulation is run in Matlab for which a mechatronic model library has been built.

2.2.3.   Collaborative Design Review and Engineering Change Management

Product Design Review (PDR) involves gathering and evaluating product design and its concrete plans for realization and improvements, so as to confirm that the process is ready to proceed to the next phase. Product design review is a typical scenario of collaborative product development. A team is tasked with the design and development of a new product. The team consists of members from multiple disciplines. Some are lead users (key customers), some are core (key) suppliers, and others may come from various functions and units of the organisation. Also, all of them are geographically dispersed.

Traditionally, design review is conducted by circulating the documents of a product design, so that they can be reviewed by one member to another. After that, a meeting is then arranged to resolve different opinions. This process is very inefficient, especially when some external members from other regions, such as key customers and suppliers are involved. Engineering Change Management (ECM) is another business process in product development, closely related to PDR process. Chapters 10 and 13 discuss ECM and PDR in more detail respectively.

Figure 2.4 illustrates the client user interface for the prototype supporting the engineering change process during the detailed design stage of product development. Note that everything is done electronically with no need to hold group meetings or refer to paper drawings. Also, the review and approval process can take place in a collaborative environment through the web site and the conferencing tools such as NetMeeting and Conference, which will be discussed in the next scenario.

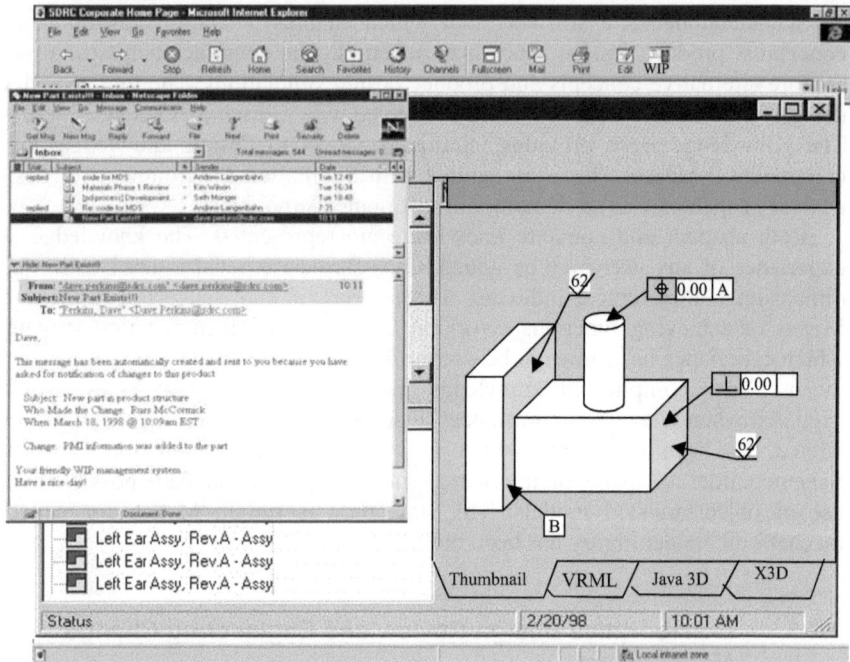

Figure 2.4 Web applications for Collaborative Design Review.
Source: Rezayat (2000)

2.2.4. Supplier Chain Integration

Suppliers frequently possess vital product and process technology that can lead to improvements in the new product development (NPD) process. Therefore, there is a need to collaborate very closely with the suppliers at this stage of product development. It is widely accepted that ESI (Early Supplier Involvement) is beneficial to both the buyers and suppliers. Typical benefits include: reduced development costs, early availability of prototypes, standardisation of components, visibility of the cost-performance tradeoff, consistency between design and supplier's process capabilities, reduced engineering changes, higher quality and fewer defects, availability of detailed process data, reduced time to market, and early identification of technical problems, etc.

In fact, ESI in NPD is extremely complicated, requiring in-depth investigations. Chapter 11 will provide a more comprehensive discussion on this topic. Here is a simple description of a scenario, in which business partners integrate their activities in the extended enterprise. Initially, the end product manufacturer prepares the design of the new product and provides the relevant design and business documents on the web site. Bids are then invited from low tier suppliers. Interested suppliers can access the web site to obtain necessary documents, so that bids can be prepared and later submitted. If necessary, suppliers can use the

NetMeeting to hold a live collaborative discussion session with the end product manufacturer. The White Board facility is brought up and everyone starts the mark-up process as appropriate. Comments from chat sessions and mark-ups from white board are all saved in the web site. The information is dynamically converted and displayed in the best possible multimedia Web format. Supplier questions regarding a particular feature may be resolved by direct interactions through the web site. Conflicts between the designers, shop-floor personnel, and the suppliers are all resolved through the web site.

Figure 2.5 shows an image of a live net-conferencing session between an imaginary OEM (Original Equipment Manufacturer) and a supplier, who may be thousands of miles apart physically. It is important to note that the whole process takes place electronically, with no need for expensive paper drawings. Notice the presence of audio and video equipment, and the ability to mark up and time-stamp the documents electronically. The discussions from each session, stored in a chronicle order in the same file, can be retrieved very easily and the complete file can be stored in the system.

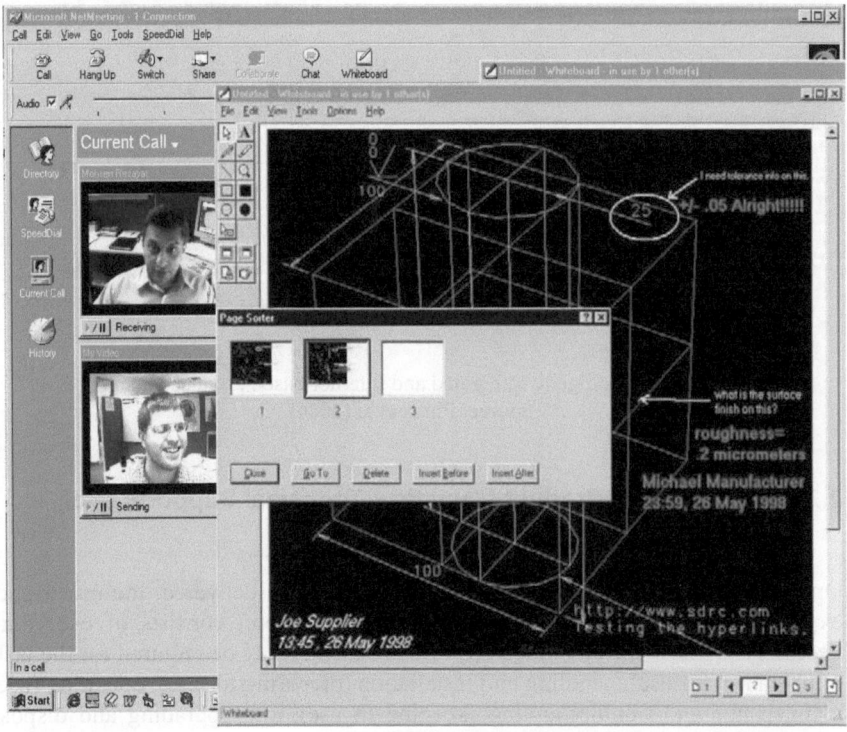

Figure 2.5 Supplier involvement and selection in new product development.
Source: Rezayat (2000)

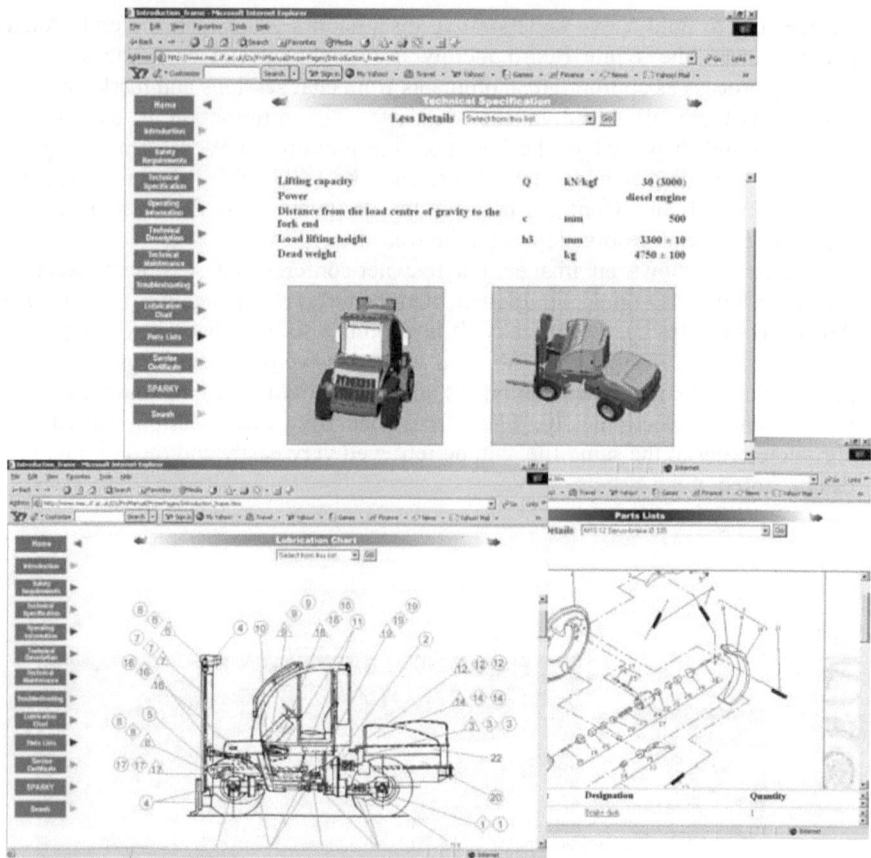

Figure 2.6 Intelligent, Integrated and Internet-based Product Manuals.
Source: Pham et al (2000)

2.2.5.   Intelligent Product Manuals – Technical Supports and Customer Services

Once a product has been designed, manufactured and delivered, the user rightly expects that it be properly supported. Product support consists of everything necessary to allow the continued use of a product. It may be required for the tasks of planning (for use), handling and installation (preparing for use), operation (use), maintenance and troubleshooting (keeping in use), and upgrading and disposal (changing and ending use). To accomplish these support tasks, the product is brought together with the necessary supplies (consumables and spare parts), equipment (tools and facilities), persons (suitably skilled) and information.

A product support network provides for the production or acquisition, storage and supply of the above- mentioned support items. It can include product training, technical documentation, help lines, servicing, spare parts ordering and maintenance management. Conventionally, all support items are brought together physically with the product while information and persons are supplied remotely, e.g. by telephone link. Despite this provision, product support can still be costly, labour intensive and of poor quality, from both the supplier's and user's point of view.

Intelligent product manuals (IPMs) are designed to supply the user with the product information of such high quality that the task of the user is effectively de-skilled. Figure 2.6 shows an example of IPM. Thus, the product becomes easy to use and maintain by the virtue of this enhanced task support. The benefits of this type of system are reduced need for skilled persons and for training new technical staff (decreased cost) respectively, and better and quicker task performance (reduced cost and improved performance). Enhanced electronic communication between the hardware, the information systems and persons involved in product support creates some other opportunities to be considered alongside with IPMs. These include computer-based training, remote hardware monitoring (e.g. via the Internet), tele-presence of skilled persons (e.g. by video links) and integrated spare parts ordering and maintenance management systems.

2.2.6. Collaborative Product Development Project Management

Collaborative Product Development (CPD) has been an area for intensive research for two decades. Certain success factors are: teamwork; better communication, project management, information sharing and consistency. Figure 2.7 shows an overview of a prototype web-based framework, called POPIM (Pragmatic Online Project Information Management), for managing collaborative product development projects within an extended enterprise environment (Huang, Feng and Mak, 2001). The framework provides a common workspace for geographically dispersed project team members to communicate, share, and collaborate on a project through online access to the most up-to-date project information. As a result, a high-level data consistency can be maintained, and experience and insights can be accumulated to form the knowledgebase. In addition to standard project management functionality, such as defining work structure breakdowns, determining work schedules, teaming up with specialists, and allocating resources, POPIM incorporates workflow management (including dependency management), and deliverable management (document management if documents are considered as one kind of deliverables). Individual members have their personalized accounts according to their skills and roles/responsibilities in a project. A project team and its members may maintain their own journals/records. More application-specific functions, such as product design review and engineering change management can be implicitly performed through online document forms.

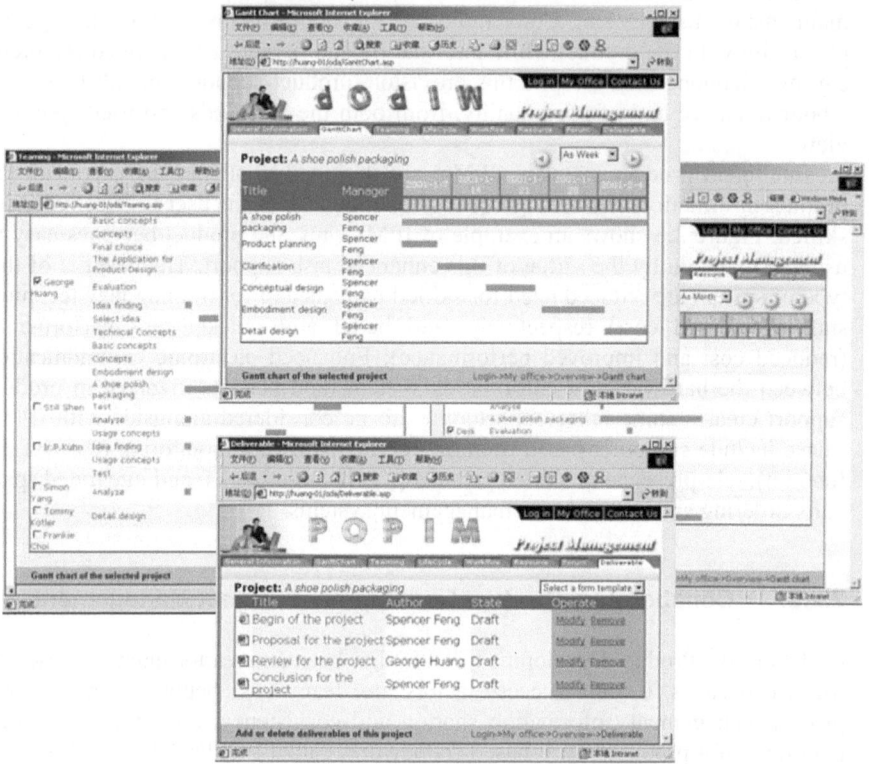

Figure 2.7 Managing product development project over the web.
Source: Huang, Feng and Mak (2001)

2.3.   SUMMARY

The World Wide Web is gaining increasing attention in developing product design and manufacturing applications on the Internet/intranets. One attraction is its inheritance of the client/server architecture for distributed computing. This is well suited for resolving many difficulties in collaborative product development, which involves team members and computerised tools distributed geographically and over time. Furthermore, the web browsers provide an open standard user interface for accessing remote services. Web applications require no installation, maintenance, configuration, or upgrading on the client side if implemented and deployed properly. Therefore, the users need not to deploy extra resources for these activities while they can still enjoy the latest services on the Internet/intranets.

The web technology has found wide applications in product design and manufacture. The literature has indicated several main areas of web applications.

Many individual decision support systems are expected to emerge through the web. Such systems seem to exist at every stage of the product development cycle, ranging from the customer requirement analysis through the design and process planning to the post-manufacturing after-sales services. However, the interoperation between these web applications remains to be a main issue to be resolved in the near future. Although this interoperation aspect is closely related to the development of web-based teamwork and collaboration frameworks, it is yet to be investigated whether such interoperation and collaboration should be better achieved between the application servers or on the client side. Finally, special-purposed search engines appear to be expected to support the use of web applications in product design and manufacture. This would help the practitioners to find and apply the most suitable tools easily and quickly on the Internet/intranets.

Despite all the efforts made in developing web applications so far and potentials that the web technology seems to promise in early explorations, the web-based approach is just in its infancy.

Although the web technology sounds highly relevant and promising, it is still not clear exactly how effective it can be used to support collaborative product design and manufacture, and what kind of extensions should be introduced to its present form. Although allowing multiple, simultaneous, and distributed accesses, web pages providing hyperlinks leading the user to jump freely anywhere they point to, result in some difficulties in providing better workflow management required in collaborative product development. In addition, it requires intensive investigation into how best to provide and utilise, develop and deploy web applications in product design and manufacture on the Internet. The next chapter aims to present some general guidelines in this respect.

3

CHALLENGING ISSUES

The impact of the e-commerce/e-Business market force has initiated rapid development in terms of the technology advancement and public/professional awareness. A pace of rapid progress in practice can be achieved and sustained only if the "inertia" issues are resolved surrounding the development of web applications, deployment of web applications to form enterprise portals, and operation of web applications to support business activities. This chapter attempts to identify some of these limiting issues while their solutions are yet to be explored by both the research and practitioner communities in the near future. Some of the questions that will be addressed in this chapter include:

- Are changes in business operations necessary?
- When and where web applications are applied most effectively and efficiently?
- Computation shared between the server and the client
- Interactivity and security
- Convert existing legacy systems into web-based on the internet
- Choose appropriate web applications
- Decision traceability
- Individual versus common working memories
- Collaborative workflow management
- Synchronous versus asynchronous
- Difficulties in web application development
- Business models of digital enterprises

- Information overflow
- Enterprise portal servers

This chapter discusses these issues under the four general categories. Firstly, the challenging issues related to the operation of web applications are discussed in Section 3.1. Sections 3.2 and 3.3 highlight some of the key issues related to the deployment and development of web applications respectively. Section 3.4 explains some of the special aspects of web applications and enterprise portals for product design and manufacture.

3.1.   CHALLENGES RELATED TO OPERATION OF WEB APPLICATIONS

3.1.1.   When and Where Web Applications are Applied Most Effectively and Efficiently?

Researchers and practitioners have generally been uncertain whether or not web applications are suitable for product design and manufacturing. For example, the following conditions have been identified in the Virtual Manufacturing initiative regarding when the web technology is most effective (http://www.isr.umd.edu/Labs/CIM/):

- Interactions between the user and the remote servers should not be very frequent, and should not require a very high bandwidth.
- Web-based tools must be able to serve a large group of users, but relatively difficult or costly to get copies for direct use at user sites with respect to how often any single user might want to use them.

The above two conditions were identified some time ago. Their validity needs further examination at the present situation after rapid development of the web technology over the past few years.

Let us have a look at the first condition first. Many product development activities require frequent interactions between the engineers and the systems. The restrictions imposed by the above first condition are actually diminishing with the development of the web technology. For example, frequent user-system interactions can be broken down into two phases: between the user and the client-side system and between the client and the remote server. While the interactions between the server and the client machines are kept at minimum levels through careful allocation of computation among the server and client, high interactivity between the user and the client machine is achieved through client side scripting and processing.

The second condition is not unique to web applications. Indeed, majority of computer application systems is developed with a user group in mind. Serving a wide range of users at the lowest cost is a notable advantage of web applications rather than a restriction or condition.

In reality, the spectrum of web applications in the literature as discussed in Chapter 2 seems to suggest that the web applications have spread over most aspects of product development without restrictions from the above two conditions. The future development is highly likely to disregard these two restrictions.

3.1.2.   How to Choose Appropriate Web Applications?

Hundreds of web applications in product design and manufacture are expected to become available on the Internet in the world. The question is which one is the most appropriate for a problem at hand. This is a practical question and difficult to address. Currently, designers and manufacturing engineers spend more than 50% of their time just looking for information. Surfing on the Internet without clear criteria or objectives is time-consuming with general-purpose search engines. On the one hand, the system developers would or should promote the awareness of their tools. On the other hand, the users would like to find a tool quickly and easily. In fact, both are not easy to achieve on the Internet at present.

Once located, a shared web page can be created to provide hyperlinks to individual web-based tools. The user can follow the logic built into this shared page, from one page to another. This certainly is a compromise between some and none workflow management. It is still not sufficient for some simple reasons. The next chapter proposes using the concept of Virtual design Office to register and provide web-based design services. Such Virtual Design Office is basically a special-purpose search engine.

ACORN (Advanced Collaborative Open Resource Network) is a major project under the Agile Manufacturing Information Infrastructure (AMII) program. The aim is to create a scalable prototype of a national (USA) network of design and manufacturing services. Such a network would revolutionize product development by encouraging widespread design re-use, analysis and simulation. A critical part in creating the ACORN infrastructure is developing the community that will use and expand the network of services.

3.1.3.   Appropriate Business Models

Web applications operate in a different way from standalone software applications. Furthermore, web applications have fundamental impacts on the business models of enterprises that use them. Unfortunately, the research into business models lags far behind the development of the digital enterprises. It is not quite clear what is the most appropriate business model, especially in the context of manufacturing industry. Further in-depth investigations are envisaged to take place in the following two directions:

- Business models for marketing, deploying and/or licensing web applications, i.e. business models of application service providers.
- Business models for manufacturing enterprises that employ the web applications extensively, i.e. business models of digital manufacturing.

At present, it is not clear if the business model of a manufacturing enterprise needs to be changed when the digital e-business approach is adopted. In the promotion literature, it becomes a common slogan now to promote e-commerce/e-business: "The Internet and web technology is changing the way that the business is done and the way that we live." Indeed, many e-business software systems do require certain changes in business operations, that is the ways that we work. This has proven the hardest in introducing and implementing any new technologies or initiatives, no matter how potentially promising they are. Digital enterprise is no exception.

The worry about the changes in business operations is understandable given the fact that the web technology itself is evolving extremely rapidly. This kind of catch 22 cycle can only be improved through proven benefits achieved by pioneering organizations.

3.1.4.   Decision Traceability

The web technology does not provide sufficient supports for decision traceability, especially when multiple web-based decision support systems are employed in a project. Let us assume a simplistic situation where each web-based decision support system is accessed through one web page. The user uses the web browser to connect to individual systems one after another. After connection, the user enters input data using the browser and submits them to the corresponding server for decision-making. The result is returned to the same web page (usually a separate page). After the first decision support system finishes its task, the second system is activated in the web browser. Some of the input data to this second system may come from the output results from the first system. The previous web page for the first system may have lost its data. This makes it difficult for the user to use the second system.

Product development is highly iterative. For example, the third decision support system may prove some of the previous decisions made in the first and second systems are not good enough. The user has to return back to the first and second systems to generate new decisions. The user will probably have to modify some of the input data while keeping the others. The web technology at present is very weak in this capability.

Web pages can be designed to cache data inputs in previous pages so that they are retained when the user goes back to these pages. This can only be achieved in the same session. However, it is sometimes necessary for the user to quit the session because of natural requirements such as tea breaks. When the user returns to work, the user starts a new session. However, all the data inputted or generated in previous sessions of the same project are not maintained. This causes significant problems in continuing the task.

3.1.5. Individual versus Common Working Memories

Individual web-based design tools are usually developed by different third parties independent of each other. They have their own private working memories that cannot usually be shared by others. However, when multiple web-based design tools are brought together for collaborative product development, it is usually required to transfer information between these systems at certain levels. In addition, such information should usually be kept for future reference. It is impossible to achieve this with private working memories for a collaborative product development project. Therefore, a shared common working memory may be necessary for individual tools to share common data. In addition, the shared common memory can also act as a media for communications between individual web-based tools.

The metaphor of blackboard has been introduced for computer aided collaborative problem solving (Nii, 1986). In fact, it has also been used to support computer aided collaborative product development (Laliberty et al, 1996). Sending and getting data to and from the common working memory are an issue for further investigation.

3.1.6. Collaborative Workflow Management

It is interesting to note that the web technology on one hand has been used to develop web-based supports for group work and workflow management, and on the other hand it is suffering an inherent limitation in workflow management itself. The web technology has been applied to develop web-based collaborative workflow management systems. GS_{Web} (GroupSystem for the Web) provides an environment for group coordination and a suite of collaborative tools (Romano et al, 1998). WebWork is a fully distributed workflow enactment system relying solely on the web technology (Miller et al, 1997). These are web-based supports for collaborative group work where humans are the decision makers and their computers assist them in better communications. They do not support such scenarios where decision makers may be web-based design tools.

Web users easily experience frustrations in loosing the focus and direction during the navigation. Hyperlinks of web pages can lead the user to jump from one page to another. After a few jumps, the user may become unclear where he/she is, from where he/she comes, and where to proceed. At present, browsers provide navigating facilities, going backwards or forwards or jumping to specific web pages. The flow of these web pages does not reflect the flow of the work involved in the collaborative product development project. One compromise is to build certain logic into the shared web pages where the user can always return. This simply sends the user to the starting point but losing the flow of logic.

In a specific product development project, some activities must be carried out after other activities have been completed. Such dependencies cannot be easily reflected in the current web technology. The above brief discussion simply implies the need for a workflow navigator. It will deal with decision traceability, common working memory, and workflow management.

3.1.7.   Synchronous versus Asynchronous

It is highly expected that the web applications are able to support distributed teamwork. The geographical dimension of distribution is adequately addressed because remote users may use web browsers to access relevant web applications on the Internet. However, the time dimension is still hard to deal with. Let us have a look at two typical scenarios. One is that team members work on a project with the support of a web application on the Internet in the synchronous mode. They share the same workspace for deriving and presenting their decisions. They interact with each other just as they use the same standalone application at the same location. Decisions made by one participant are reflected in all the clients. The other typical scenario is that one or two members cannot attend such synchronous teamwork because of other engagements. The question here is how such members can be involved later in the teamwork. One solution is to incorporate the replay facilities in such web applications. Absent members may replay the previous synchronous collaboration session and decide if there are any opportunities for participating. Although asynchronous, this kind of participation is still better than complete exclusion.

In computational terms, synchronous and asynchronous web applications are now not too difficult to implement with the current Internet and web technologies. A main issue is that the application clients in this case must also act as servers for the central application server. That is, a decision made by one client is first sent to the server. The server records the decision in appropriate format and then notifies all the other active clients that must perform some tasks to update their own workspace locally.

The main difficulty in developing and applying synchronous applications is the resolution of conflicts between simultaneous users. For example, one client may attempt to delete a previous decision while another may plan to extend the decision further. If it is deleted, then it cannot be extended any further. Another example is that a decision made by one client may not be acceptable to other users. They have to resolve their differences of opinion soon or later. The issue of conflict resolution has been discussed in the context of collaborative product development (Klein, 1992). The challenge lies in how the proposed strategies should be best incorporated in synchronous web applications.

3.2.   CHALLENGES RELATED TO DEVELOPMENT OF WEB APPLICATIONS

3.2.1.   Difficulties in Web Application Development

Because of the technological complexity, the web application development varies widely from the development of traditional software application systems.

While the popularity of the web and its advantages as a client-server platform have led to countless HTML-based applications, the development of web applications is still mostly ad hoc. There is not rigorous, systematic approach, and most current web application development and management practices rely heavily on the knowledge and experience of individual developers. One main reason is that the web is regarded as an information medium rather than an application platform. The development of web applications is therefore seen primarily as an authoring problem rather than a software development problem to which well-established software engineering principles should apply.

3.2.2.   How to Share and Distribute Computation between the Server and the Client?

In web applications, computation is distributed and shared between the server and the client on the Internet. Generally speaking, servers are powerful machines and therefore should be used for tasks involving more number crunching. On the other hand, local clients are less powerful but with good Internet connectivity and therefore should be mainly for checking input and output constraints.

If all the computation is carried out by the server component, then the client must submit every bit of the user input to the server. Clients and servers communicate to each other through HTTP by exchanging HTML files. This process is in many cases more time consuming than data processing by either the server or the client. Because of this limitation, it becomes problematic to maintain a reasonable level of interactivity between the client and the user. That is, the user often has to wait for the client to contact the server and receive HTML files. This is usually undesirable in many product development activities where high interactivity is necessary between the user and the computer aids.

One way to minimize the communication with servers is to let the client to accomplish the computation instead. This leads to more sophisticated clients that usually take longer to download when the web page is accessed, as well as complicating the deployment in some cases.

In order to maximise the computational efficiency, care should be taken when designing web-based applications, e.g. whether a computational component should be deployed on the server side or client side and when communication between the server and the client should take place.

3.2.3.   Interactivity and Security

As already discussed above, the communication between the server and the client machines is expensive. This implies that the server-client interactivity is limited. On the other hand, frequent interactions are usually necessary between the human user and system in collaborative product development. This high interactivity requirement can be achieved through a combination of proper allocation of computation between the server and client and speedy interaction of the client with the user enabled by client-side scripting and processing. The ActiveX technology provides a number of ways of client side scripting, e.g. VBScript, standard ActiveX controls, ActiveX code components deployed on the client side, and ActiveX documents, and/or Java equivalent. These techniques and facilities enable the developers to develop web-based applications that are as equally interactive as any other standalone computer aided systems.

However, HTML and client-side scripts are interpretive, instead of being compiled, from source codes. This raises the concern about the security of the source codes. First, web-based design tools are usually copyrighted intellectual properties and should not be reproduced without consent from the copyright holder's permission. Client side VB or Java scripts are usually made accessible to all the users at the client site. Anybody would be able to cut and paste the scripts. In addition, client side scripts are usually downloaded automatically from the server during the access. It is necessary to protect the client machines from Internet vandalism. This is why VB scripts and Java scripts are designed not to be able to damage the client machines. For example, data access to the local clients is not possible but usually necessary in collaborative product development. The ActiveX technology allows such data access, and in fact the full functionality same as standalone systems on the local machines. Unless the server source is trustworthy, damages are likely to be made by ActiveX codes to the client machines.

In addition to the above worry about the security from the software vendors, another security worry is from the users. Two aspects are important. One is how the service providers ensure the confidentiality of the user information, both individual and corporate users. Another is how the software and service providers enforce the access authority. The users must be assured before they commit to the frequent usage of the web applications.

3.2.4.   How to Convert Existing Legacy Systems into Web-Based Systemson the Internet?

Over the last two or three decades, a large number of standalone computer systems have been developed and applied in industries. To make things even worse, some manufacturers have just been persuaded to invest in IT/IS (Information Technology and Information Systems) and now these technologies are made obsolete by the Internet and web technology. Now the challenge lies in how to convert these legacy systems into web-based counterparts. One solution is to recode the systems completely in an Internet programming environment. The newly converted systems are accessible on the Internet/Intranet through the web browsers. But there are a number of limitations associated with this approach. First, re-coding is time-consuming. Second, it would be difficult to fully understand the logic of the existing legacy systems especially after the original developers/programmers have left.

An alternative solution to the conversion problem is to wrap the legacy systems in an Internet programming environment. In order to overcome the above limitations, efforts have been made to encapsulate the existing legacy systems so that they can be made accessible on the Internet through the Web browsers. Wrapping templates can be developed. One advantage of this approach is that the partial re-coding effort is kept at its minimum.

3.2.5. Information Overflow

The ultimate purpose and strength of the web applications is to provide and process the right information when it is needed in the right way. This strength may defeat its purpose if the web applications are not properly designed. One consequence is the overflow of information.

By information overflow it is meant that excessive amount of information is made available for the user's purpose in the inappropriate format. System developers are often confused between the requirement for teamwork and information overflow. To solve a real-life engineering problem such as those in product design and manufacture, it is not rare now for a web application to require multiple participants from the project team to contribute input data and expertise, and to produce meaningful output information. This is not however an excuse to squeeze data and information of differing categories into one user interface or web page as has been done by naïve developers and can be seen from some commercial systems. Such a design would require more than one member to deal with one web page. An alternative to resolve this type of information overflow is to categorize the information according to the user's background and responsibility and then incorporate them into one or related web pages. By so doing, one user is able to understand and handle all the information on the web pages that they are supposed to handle without affecting and involving other members.

3.3. CHALLENGES RELATED TO DEPLOYMENT OF WEB APPLICATIONS

3.3.1. Difficulties in Deploying Web Applications

Significant difficulties exist because of substantial inherent disparities among web applications that constitute an enterprise portal:

- Web applications are usually supplied by third parties, sometimes competitors.
- Web applications may be developed and implemented under different environments with different constructs.
- Web applications may require specific hardware platforms and operating systems for deployment.
- Web applications may have different levels of security and access control.
- Web applications have their own data sources although it is possible for them to share a common repository.

3.3.2.   Approaches to Building Enterprise Portals

There are at least three approaches to building enterprise portals and their features are summarized in Table 3.1. In a recent enterprise portal market study, Meta Group found that 75% of enterprise portal development efforts where still being accomplished using labor-intensive traditional web development environments. Their research concluded that companies should use portal frameworks or portalets, rather than conventional web tools to build their enterprise portals.

"Applets" and "Servlets" are well-known Java constructs for building up client and server components of web applications, respectively. Similarly, the term "Portalets" is introduced here as basic constructs of software solutions for creating enterprise portals from enterprise web applications. This introduction is necessary when more and more web applications are put into practice to support business operations, leading to the emergence of digital enterprise. The scope of this paper is limited to the manufacturing context, especially product design and manufacture.

3.3.3.   Enterprise Portal Servers and Enterprise "Portalets"

Portalets themselves are web applications for managing other domain-specific web applications. As a matter of fact, some utility web applications such as calendar, email, online chatting, bulletin board, whiteboard, etc. are delivered as constituent components of some commercial enterprise portals.

Enterprise portalets range from simple Intranet indexing and search tools to enterprise-level database-driven information storage and retrieval products. Table 3 summarizes, without prejudice or preference, some of the typical enterprise portalets commercially available on the market. Although they offer competing features and unique facilities, they serve the similar purpose of facilitating the deployment of web applications to form enterprise portals. Figure 2.1 in Chapter 2 shows the architecture of the appsolut Enterprise Portal Suite.

An enterprise portalet differs from general-purpose website development environments such as Microsoft Frontpage while sharing numerous common functions. A website development environment is generally considered as an environment for developing and deploying individual web applications ranging from simple HTML contents to web pages with sophisticated computational components embedded or attached.

In contrast, an enterprise portalet provides an environment for deploying various individual web applications into a central website called an enterprise portal. There is no doubt that environments like MS Frontpage have been used for this purpose in practice. However, this approach is unlikely to meet the requirements set out for many commercial enterprise portalets.

Table 3.1 Enterprise portal approaches.

	Out-of-the-box web applications	Build from scratch with general-purpose web tool	Portal framework or Portalets
Time to solution	Fast time to solution if requirements match vendors offering	Long time to solution through extensive, manual coding	Fast time to solution while matching specific requirements
Effort/skills required	High effort to adapt to specific requirements; vendors' consultants needed to adjust portal	High development effort, experienced, highly skilled IT people required	Average IT skills needed through reduction of complexity and componentization
Customization	Hard to customize to specific requirements; mostly domain-specific (ERP, CRM, PDM)	Design and development according to specific requirement	Design and assemble according to specific requirements
Flexibility after deployment	Hard to adapt to new requirements and backend systems	Extensive coding needed to adapt to new requirements and backend systems	Comparatively easy exchange and adaptation of components to incorporate new requirements and backend systems

After appsolut White Paper at
http://herkules.appsolut.com/eps/en/PDFs/Whitepaper%20A4.PDF

3.3.4.   Types of Enterprise Portalets

Enterprise portalets can be grouped into the following three categories:

- Anchor Portalets. An enterprise portalet of the "Anchor" type provides a central and/or federated data repository around which web applications are installed.
- Plug-in Portalets. An enterprise portalet of the "plug-in" type manages enterprise web applications that have their own repositories, instead of sharing the central one.

- Value-added Portalets. An enterprise portalet of this type is very similar to plug-in portalets, except that it usually involves web applications other than enterprise web applications.

3.4.   SPECIAL REQUIREMENTS OF CPC WEB APPLICATIONS

As this book is mainly concerned with web applications in product design and manufacture, we use the term Collaborative Product Commerce (CPC) web applications. In addition to those issues already discussed in a previous section concerning the difficulties of deploying the general-purpose web applications, there are issues specific to CPC. CPC web applications have their own unique characteristics compared with other types of enterprise web application. For example, unlike e-Commerce, some web applications in product development and realization are concerned with designing new products from customer requirements, and therefore the products do not exist yet. The transactions between players are far more than merely data and information in popular e-commerce. Sophisticated knowledge are involved in the transactions. The variety and intensity of transaction information are far more sophisticated than those in popular e-Commerce. For example, 3D display and manipulation of geometrical information of products and processes on the Internet through the web remains a great challenge. The frequency and intensity of negotiation and collaboration between team participants within an enterprise (design, manufacturing, assembly, marketing, management, etc.) and/or across different enterprises (business partners, suppliers, customers, etc.) are far greater than those in popular e-Commerce processes.

Therefore, the CPC portalets must be able to accommodate these features such as 3D display and manipulation of geometrical information of products and processes on the Internet. In addition, CPC web applications usually involve project-oriented. A project usually involves a number of work packages or tasks which are assigned to several members of the project team. Each member may use a specific web application. Because work packages or tasks of the same project are closely related, their corresponding web applications must interact intensely. Although special web applications may be introduced to enhance team-based collaborative project management, it remains a great challenge to make CPC web application more interactive with the users and between themselves.

3.5.   SUMMARY

This chapter has highlighted some of the major forward-looking issues that must be resolved if web applications and enterprise portals are to play effective and efficient roles in product design and manufacturing. At present, these issues have not yet been received serious consideration when developing and applying web applications and enterprise portals. This is because most of the research projects are at their earlier stages where focuses are placed on functionality and implementation issues. Although advanced developers and practitioners will become aware of and are forced to address theses issues at later stages, this chapter argues that the issues should be addressed as early as possible. The reason is the same as any other product development projects where late changes are expensive and time-consuming to make.

Having said that the issues must be tackled as early as possible, it must be pointed out that there are no firm solutions so far for the many challenging issues discussed in this chapter. Developers have to rely on their experiences in designing, implementing and deploying web applications and enterprise portals. However, we must be optimistic that sufficient insights will be gained in the near future and these issues will be successfully addressed.

4

DEVELOPMENT AND DEPLOYMENT OF WEB APPLICATIONS

There have been a considerable amount of interests in providing the state-of-the-art product design and manufacture techniques and methods accessible on the Internet from web browsers. This chapter concerns with the process of building web applications for product design and manufacture. The focus is on how such systems are being developed and implemented. However, there have been considerable difficulties in the design, development, and implementation of web applications in product design and manufacture. The complexity can be understood in several ways. For example, there is no rigorous and systematic approach to the development of web applications at present. Current web application development and management practices are mostly ad hoc, heavily dependent on the knowledge and experience of individual developers.

Furthermore, the World Wide Web or simply web was originally developed as an information medium for distributed research teams to share documents in electronic forms. In contrast, product design and manufacture applications involve both intensive computation and high interactivity with the users. Web pages have evolved from being simply static through active to interactive. Also, the technology used to achieve interactive web pages has become very complicated.

The pace of the web technology evolution is so fast that it is hard for system developers to keep the same pace. The variety of techniques, even of the same web technology is wide and many of them achieve similar functions. This is also true between competing technologies that provide comparable features with different terminology. Even those who believed that the development of web applications is merely a matter of programming using an Internet language have found it difficult to choose the most appropriate features. Not to mention the product design and manufacture researchers who spend most of their time and efforts in developing contemporary methodologies, rather than investigating the web features. Due to the complexity, there are many typical puzzling questions. Some of them are:

- Would it be advantageous to employ the web to develop my application over standalone approach?
- Which programming and hardware environments should I choose to develop my web application?
- Which web browsers do my potential users use to run my web application and on which hardware platform?
- How best are application components deployed across the network?
- How should I develop the database for my web application?
- How should the application database be deployed on the network and how the remote database is connected and accessed?

These questions must be adequately answered before developing and implementing a web application take place seriously. Unfortunately, there are not sufficient guidelines in the literature. This chapter aims to address these questions and hopefully provide some insights based on our own experiences in web applications. The discussion starts with explaining the so-called 3-tier client/server architecture for web applications, and its differences and similarities with the usual client/server distributed computing. The two most popular web technologies, namely ActiveX and Java, are then discussed in terms of their main constructs for developing web applications. The methods for the deployment of web application clients and servers are highlighted. Remote database access and connection deserve further clarification because many web applications in product design and manufacture involve product, process and resource data.

The tenor of the discussion is discussive and tutorial. In addition, the discussion is somehow idiosyncratic, heavily conditioned by the authors' experience in this field. It is hoped that researchers are able to develop their proof-of-the-concept prototype web applications to demonstrate their methodologies after considering these issues discussed in this chapter.

4.1.   DEVELOPMENT ARCHITECTURE

The design, development, deployment and database access of web applications are dominated by the so-called three-tier architecture, as shown in Figure 4.1. Let us assume that the web application is a prototype synchronised QFD (Quality Function Deployment) system (See Huang and Mak, 1999c for more discussions on QFD). Similar architecture applies to other web applications with variations in terms of remote database connection and component deployment. The subsequent sections will provide further discussions on the variations.

The first tier, usually referred as the client tier, includes the client component of the QFD web application. When used by QFD analysts, web browsers automatically become the QFD clients. The client component is downloaded onto the client computers and executed within the client web browsers. QFD clients are not actually part of the web-based QFD system before they connect to the QFD web server. They become so only when they are connected to the QFD web server. In theory, there can be unlimited numbers of QFD clients using the services provided by the QFD web server. They may use the system on the same project or different projects.

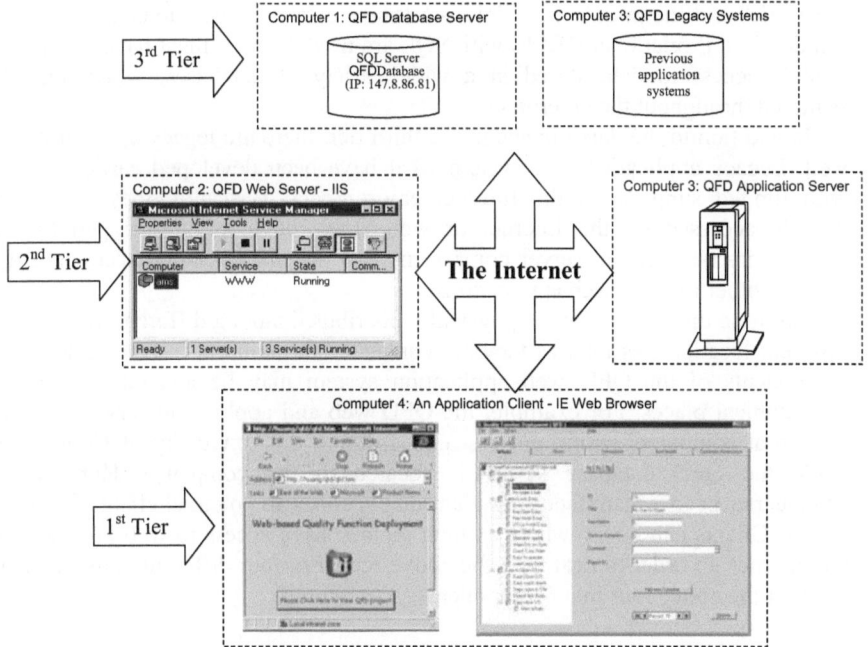

Figure 4.1 Three-tiered architecture for web applications.

There are two types of middle tier, which is often called the middleware tier. The first type of the middle tier is the QFD web server from which the users can download the application client components through web browsers, and/or the QFD application server with whose components the client components are associated. A web site is created for the QFD system at this web server, which actually includes web sites for other applications. It receives the client-side request, and processes the data, if necessary. The QFD web site includes a number of web pages to which QFD functional components are attached.

The second type of the middle tier is the QFD application server. In Figure 4.1, the QFD application server is deployed on a computer different from the QFD web server machine. The QFD web server machine is where the QFD application clients are downloaded to the client machines. The QFD web server is not involved in any further computation or data management. In contrast, the QFD application server is run on the machine where the server is deployed. The QFD application server is a piece of software developed and deployed with several responsibilities, such as for dealing with the QFD database server, accomplishing the synchronization activities and other computational activities.

The third tier, normally named the data tier, is the QFD database server. The relevant QFD data is managed by a QFD database server. The relationship between the middle tier and the 3rd tier is more straightforward than that between the 1st tier and the middle tier. This is because database operations are relatively standardised if relational DBMS with SQL are used. Servers must have secure and reliable access to data stored in a wide variety of databases, which may be scattered throughout the enterprise.

In addition to the data objects in the third tier, there are legacy applications as well. Legacy applications are systems that have been developed earlier with or with limited supports in the Internet environment. Basically, they cannot be directly accessed on the Internet or web. They have to be wrapped by the middleware (application server components) so that they become accessible just like any other Internet/web applications.

All these three tiers may be physically distributed among different machines at different locations, as long as they are connected through the Internet. Also, four components of the QFD web application system may be located in different geographical places. For example, the QFD web and application servers may be located at computers within our research group at the University of Hong Kong while the QFD database server is located in the company's Hong Kong headquarter or in a manufacturing plant that is closer to potential clients for quick access. Clients can be anywhere as long as they have access to web browsers on the Internet. In order to deal with the above scenario, four different computers are used for the above four main components.

4.2.   IMPLEMENTATION CONSTRUCTS

Web applications can be implemented in many programming languages and environments. Two of the most popular and competing choices are the ActiveX technology and the Java technology. They do not only provide specifications and object models for distributed computing but also act as programming environments. The web programming is still evolving rapidly. A number of competing web browsers such as Internet Explorer, Netscape, MOSAIC, etc. are already extended to support Java and VBScript programming.

4.2.1. HTML, VRML, and XML

Early web applications are simply static web pages compiled using the HTML. Limited facilities such as buttons, textboxes, radio buttons, option buttons, combo boxes, etc. are provided to enhance the interactivity between the user and the web page, and thus the web browser with the web server.

Because of various obvious limitations with the HTML, extensions are standardized in several directions. For example, dynamic HTML or DHTML adds more flexibility and interactivity to web pages. Active Server Pages (ASP) is able to combine the JavaScripts or VBScripts with the HTML web pages for server side scripting. XML is the new standard that formalizes the semantics of the contents of Web files and facilitates the electronic data interchange (St. Laurent, 1998). Major advancements are expected in the near future in this direction of using XML for web applications.

Web applications in product design and manufacture are unique in that geometric information is usually involved. As it is well known that the HTML is limited in its capabilities and flexibilities, VRML files are often used to enhance the web pages. VRML is an Internet standard for communicating "rich" 3D data. The next generation of VRML is the XML-based Extensible 3D or X3D. The richness of VRML is due to its interaction and navigation capabilities, plus the fact that VRML objects can be hyperlinked to multimedia (image, text, video, audio) or HTML files, as well as to other VRML objects. Many CAD systems such as SDRC's I-DEAS currently have the ability to create such outstanding VRML files (http://www.sdrc.com/ideas/vrml_help/vrml_help.html).

4.2.2. ActiveX Technology

The Microsoft ActiveX technology allows programmers to easily assemble reusable software components into sophisticated applications and services in an intranet/Internet environment with minimum efforts (Swank et al, 1997). These reusable software components are called ActiveX components. They are existing program codes and data made of one or more objects using the ActiveX technology. The ActiveX technology also intends to encapsulate Java. In fact, ActiveX components can be created by using Microsoft Visual Studio toolkits such as Visual Basic, Visual Java ++, and Visual C++.

Several types of ActiveX components are available. They include ActiveX-enabled applications such as Microsoft Office; ActiveX controls are standard user interface elements that allow the user to assemble forms and dialog boxes rapidly; ActiveX documents, and ActiveX code components. They can be used to develop both application servers and clients. ActiveX components can be used to develop applications that integrate tightly with the other elements of the Internet or Intranet site.

Application clients can be compiled into ActiveX controls, then embedded into HTML web pages, or into ActiveX documents that are attached to HTML web pages but are usually executed in a separate container. The downloading, installation and execution processes of both ActiveX controls and documents are similar. When a user accesses a URL with an ActiveX component, it is downloaded from the server and then registered on the client machine along with the HTML page that uses script to invoke the ActiveX component.

As can be seen from above discussion, ActiveX components are easier to distribute, install and maintain than standalone application clients. ActiveX components install themselves from a central code base. No information system personnel is needed to perform software installations. When a user accesses the revision via a web browser, the ActiveX component automatically upgrades itself. Because ActiveX components are run on client machines, the operation is usually faster than running applications from a networked server.

However, there are notable differences between ActiveX documents and controls. ActiveX controls are usually embedded in the HTML web pages. ActiveX documents are non-HTML documents that can be viewed and edited in a web browser. The HTML page is replaced by the ActiveX document, which executes in the web browser as its application container. ActiveX documents also offer more complex client-side processing than HTML pages with ActiveX controls. For example, ActiveX documents may have all the elements as standalone systems such as pull-down menus but ActiveX controls do not have such capability.

An ActiveX code component is an object or objects exposed by an application that can be controlled programmatically by other applications. ActiveX code components can be used to add functionality to an HTML page on the client-side just like with ActiveX controls. However, it is more often to deploy application servers as ActiveX code components. Server-side ActiveX code components can be used to customise the creation and return of an HTML page, just like CGI programs. In addition, they can be used to manage a database connection, and marshalling queries received and results returned.

4.2.3.   Java Technology

Java was originally proposed by Sun Microsystems (http://java.sun.com/). It is a robust, general-purposed, high-level programming language and a powerful software platform. It is portable, object-oriented, distributed, and multi- threaded. Java is often described as the choice of "write once, run anywhere". All Java programs must run on the Java Platform that has two components: The Java Virtual Machine (Java VM) and The Java Application Programming Interface (Java API).

With the wide variety of Java API, many types of programs have been developed. They include Java standalone applications that run directly on the Java platform, Java applets that run within a Java-enabled browser for client-side processing, and Java servlets that run within Java enabled web servers for server-side processing. A Java stand-alone application should be delivered and installed for each client before it is run from the command line while stand-alone applications can directly access local resources as well as remote servers.

Whether embedded with or attached to web pages, Java applets depend on web browsers for their installation (downloading) and execution in the client machine. After being connected to the appropriate web server page, the whole process is automatically accomplished by the web browser. The user is unaware of the activities behind. Java applets eliminate the limitations of Java applications mentioned above. No client configuration is needed and the maintenance and upgrading are automatic from the users' viewpoint.

Java servlets are modules that run inside Java-enabled web servers and provide efficient server-side programming. They also provide a way to generate dynamic documents. They are both easier to write and faster to run when compared with CGI programs. A Java servlet handles client requests through its service method. The service method supports standard HTTP (HyperText Transfer Protocol) client requests by dispatching each request to a method designed so as to handle that request.

The combination of Java applets and servlets/servers provides the most important and promising uses of Java for implementing the web applications without the limitation of platforms. Specifically, after the connection between an applet and a servlet is established, the applet-servlet communication can usually be implemented by object serialization to transfer an object from the applet to the servlet, and vice versa.

Java Beans are components built in Java. The JavaBeans specification describes new component architecture for Java to facilitate component code development and reuse. The JavaBeans component model is based on Java classes. The model can make Java classes toolable and reusable by simply adding a few rules. Beans can be deployed as servers and clients. A bean is a reusable component that can be used to create applets, applications, or even HTML pages. The same bean should be able to play in any of these containers. The emergence of the JavaBeans component model is expected to further simplify the web application development in the Java environment.

4.2.4. Summary of Implementation

Both ActiveX and Java web implementation technologies have been used in our research projects. For example, ActiveX technology has been used in developing web-based Design for X tools (See Chapter 8) and web-based design tools such as FMEA (See Chapter 9) while Java technology has been used in developing the WeBid system (See Chapter 11).

It has been found that both ActiveX and Java technologies serve our primary purpose well. That is, they are effective and efficient in implementing prototype web applications to demonstrate the feasibility and functionality of product design and manufacture techniques and methodologies.

In terms of ease of programming, it has been found that both ActiveX and Java technologies are well supported. For example, Visual Studio provides Visual Basic, Visual C++ and Visual J++ for creating ActiveX components easily. Facilities are also provided for deploying the resulting application components.

Likewise, Visual Café provides a comparable programming environment for creating Java components.

To sum up, both ActiveX and JavaBean component models provide comparable features to adequately fulfil the objective of developing proof-of-the-concept prototype systems as set out in most research projects. The final choice is really a matter of personal preference and/or previous experience. The choice for commercial development would be largely affected by the company policies. A point worth mentioning at this stage is that ActiveX is platform specific to Microsoft operating systems while Java is basically platform independent but language specific.

4.3.   DEPLOYMENT OF WEB APPLICATIONS

Following the client-server architecture, a web application usually includes two parts: the application server and the application client. It is a real challenge to determine what computational tasks should be accomplished by the 2nd tier server components and the 1st tier client components respectively. The ultimate distribution of tasks must depend on the nature of the application problems. Generally speaking, there are two typical combinations: "Fat clients + Thin servers" and "Thin clients + Fat servers". A "Fat" component is responsible for the majority of the computation, and a "Thin" component is doing little computation. For example, a HTML web page as a client may not be involved in computation at all. Instead, the web browser is only responsible for rendering HTML to display, inputting to and outputting from the server component. In contrast, a fat client may be responsible for all the computational tasks and no servers are involved at all.

When looked in detail, the client and server components of the web applications can be deployed in various ways either dependent or independent of web browsers and/or web servers. Four possibilities are as follows:

- As a standalone application on the client side only without any application server and web browsers.
- As a web browser-dependent component on the web server but without any application server.
- As a web browser-dependent component on the web server, in combination with components on the side of the web server.
- As a web browser-dependent component on the web server, in combination with application servers on the server side other than the web server.

This section is mainly concerned with how these components are deployed on the Internet at both the server and the client side, and how the two components interact with each other.

4.3.1.   Deployment of Application Clients

Clients of web applications are deployed using web browsers to various degrees. The clients can be simply HTML pages, components embedded in or attached to HTML pages, and standalone programs downloaded from a web site and then installed, configured and executed on the client machine.

HTML pages are typical clients of web applications. The standard HTML specification provides a subset of standard HTML tags and attributes for defining the forms and associated actions. Furthermore, most web browsers have been extended to support client-side scripting using VBScript or JavaScript. Therefore, some client-side computation is enabled. Typically, client-side scripts are used for checking the validity of the data inputted by the user before submitting it to the server, for accessing document object properties and methods, etc.

Client components of web applications can be implemented as Java applets, ActiveX controls or documents. Such components are usually embedded in or attached to web pages that are maintained by the web server at a web site and web browsers are used to connect to a web site. Client components associated with the web page are automatically downloaded to, installed and executed at the client machine. The whole process is automatically accomplished by the web browser after being connected to the appropriate web server page. Thus, the user is unaware of the activities behind.

Client components look after the computation at client machines. The user interacts with the system as frequently as necessary, without incurring any communication overheads with the server. No further communications are necessary with the server since there is no server at all. One disadvantage is that the downloading time may be long if the client is "fat".

Strictly speaking, standalone systems as application clients are not classified as web applications because their execution does not require a web browser. A standalone application should be delivered and installed on each client before it is run from the command line. It can directly access local resources, as well as remote servers. One of its disadvantages is that the underlying user support of maintenance and upgrading may be drastically increased due to more delivery. This can be overcome by making applications downloadable from a web page. Possible reconciliation is that the user can first download the file(s) and then follow the instructions to install the application locally. By doing this, the user can always have the up-to-date version of the application. In this case, such clients are loosely referred as web applications as web browsers are involved in installation and configuration.

However, the user may be forced to configure the client machine and the application after installation but before the application can be used. If the user does not have sufficient knowledge or experience, client-side configuration may prove troublesome. Even if the instructions are clearly written down and easy to follow, errors may still occur frequently. Also, configuration often forces the user to unwillingly restart the machine after installation.

4.3.2.   Deployment of Application Servers Dependent on Web Servers

Usually web browsers are not involved in deploying application servers. However, web servers are widely used. For example, CGI (Common Gateway Interface) is currently the predominant model for deploying application servers over the Internet. Both the web client and the web server communicate to each other by a standard HTTP, regardless of their hardware configurations and operating systems. CGI plays the role of linking client-side actions, such as database access requests to server-side reactions, database access and queries, as well as subsequent dynamic generation of HTML for data presentation. CGI is most described as a means by which a web server can communicate with applications external to the server software and have those applications perform some processing on its behalf. In essence, CGI is a server-side process that serves as a go-between for the web server and other application programs, information resources, and databases. These information resources and databases can reside on the same physical machine as the web server or on machines at some other geographical locations. CGI also provides a standard interface between the web server and external applications that are commonly referred as gateway programs, CGI programs or simply CGI scripts. This interface abstracts the details of communications between the web server and the CGI program to the point where the only detail that the application developers really need to know about the interface is how to input data into and retrieve data from their applications. Therefore, it is not necessary to know the process of communication directly with HTTP servers in order to use the programs such as database query programs.

CGI has limitations, too. It is clumsy, stateless, and very slow (Orfali and Harkey, 1998). CGI is extended in several ways. One is through the server-side scripting framework such as the Microsoft ActiveX Server Framework – part of the Microsoft IIS running on Windows NT. Active Server Pages (ASPs) are the building blocks of the ActiveX Server framework. An ASP has two parts. The HTML part is returned to the browser while the script part is processed on the web server instead of the client. The ASP's output can be static or dynamic, as HTML is sent back to the client browser for presentation. Because the ASP scripts are executed on the web server, any databases or components to which the server has access to can be used. Most notably, ActiveX components can be easily incorporated into the ASP.

Another extension made to CGI is Java servlets. A servlet is a small piece of Java code that a web server loads to handle client requests. Servlets are convenient for simple applications. They make good replacements for CGI scripts. There are a variety of ways to invoke a servlet by using the servlet URL (http://java.sun.com/products/java-server/servlets/): from a browser window, within an HTML page or an applet, and from another servlet.

The standard HTML specification provides a subset of standard HTML tags and attributes for specifying the interaction between the HTML client and the web server. The form is used to acquire data from the user and then submitted to the server to do something useful with it. The "Action" attribute is where the server action really starts. The value of the "Action" attribute specifies a URL to a CGI script, ISAPI program, IDC file, or Active Server Page that will take some actions on the user-supplied input after it is posted on the web server. The action could be accessing a database based on a user query, generating product-mailing request to a shipping department, and/or running some other applications, depending on actual applications.

4.3.3.   Deployment of Application Servers Independent of Web Servers

The deployment methods discussed so far all depend on web servers. That is, they are manipulated through the web server in terms of their execution and data input and output. However, application servers may also be deployed independent of web servers. They are deployed as distributed components or objects following certain specification standards. Distributed objects are distributed across a heterogeneous network and they inter-operate as a unified whole. These objects may be distributed to different computers throughout a network, living within their own address space outside an application, and yet appear as though they were local to an application. Three of the most popular distributed object paradigms are Microsoft's Distributed Component Object Model (DCOM), OMG's Common Object Request Broker Architecture (CORBA) and JavaSoft's Java/Remote Method Invocation (Java/RMI).

CORBA, introduced in 1991, is a specification that defines the interoperability rules between distributed objects on clients and servers. The most critical part of a CORBA system is the Object Request Broker (ORB). The ORB acts as a central Object Bus, over which, each CORBA object interacts transparently with other CORBA objects located either locally or remotely. It also provides the middleware services that shield the clients from complexities of remote communication with data servers. It is responsible for establishing a client/server relationship between components. It first receives requests from the client objects and then routes them to an appropriate server, which in turn delivers the service required back to the client though the ORB. Each CORBA server object has an interface, which exposes a set of methods. To request a service, a CORBA client acquires an object reference to a CORBA server object. The client can now make method calls on the object reference as if the CORBA server object resided in the client's address space. The ORB is responsible for finding a CORBA object's implementation, preparing it to receive requests, communicate requests to it and carry the reply back to the client. In order to ensure the compatibility between various ORBs, the CORBA specification requires the use of Internet Inter-ORB Protocols (IIOP). In essence, IIOP provides a common communication backbone between different ORBs by adding several CORBA-specific messages to TCP/IP, the common network communication protocol used by middleware. Since CORBA is just a specification, it can also be used on diverse operating system platforms from mainframes to UNIX boxes to Windows machines to handheld devices as long as there is an ORB implementation for that platform.

DCOM, introduced in 1996 by Microsoft, is like CORBA. It separates the interface from functionality by using an IDL (Interface Definition Language). DCOM supports remote objects by running on a protocol called the Object Remote Procedure Call (ORPC). A DCOM server is a body of code that is capable of serving up objects of a particular type at runtime. Each DCOM server object can support multiple interfaces, with each representing different behaviour of the object. A DCOM client calls into the exposed methods of a DCOM server by acquiring a pointer to one of the server object's interfaces. The client object then starts calling the server object's exposed methods through the acquired interface pointer as if the server object resided in the client's address space. Since the COM specification is at the binary level, it allows DCOM server components to be written in diverse programming languages like C++, Java, Object Pascal (Delphi), Visual Basic and even COBOL. As long as a platform supports COM services, DCOM can be used on that platform. DCOM is now heavily used on the Microsoft Windows platform.

Java/RMI relies on a protocol called the Java Remote Method Protocol (JRMP). Java relies heavily on Java Object Serialization, which allows objects to be marshalled (or transmitted) as a stream. Since Java Object Serialization is specific to Java, both the Java/RMI server object and the client object have to be written in Java. Each Java/RMI Server object defines an interface, which can be used to access the server object outside of the current Java Virtual Machine (JVM) and on another machine's JVM. The interface exposes a set of methods, which are indicative of the services offered by the server object. For a client to locate a server object for the first time, RMI depends on a naming mechanism called a RMI Registry that runs on the Server machine and holds information about available Server Objects. A Java/RMI client acquires an object reference to a Java/RMI server object by doing a lookup for a Server Object reference and invokes methods on the Server Object, as if the Java/RMI server object resided in the client's address space. Java/RMI server objects are named using URLs and for a client to acquire a server object reference, it should specify the URL of the server object as specifying the URL to an HTML page. Since Java/RMI relies on Java, it can be used on diverse operating system platforms from mainframes to UNIX boxes to Windows machines to handheld devices as long as there is a Java Virtual Machine (JVM) implementation for that platform.

An excellent tutorial on the comparison between DCOM, RMI and CORBA is at the following website (http://www.execpc.com/~gopalan/misc/compare.html). According the test data provided in the book (Orfali and Harkey, 1998), the applications deployed independent of web servers outperform those deployed on web servers using CGI and Java servlets.

4.3.4.   Summary of Deployment

Application systems can be deployed on either server side or client side or even distributed on both sides. How a particular application should be deployed depends on different requirements such as interactivity, integrity, and speed, etc.

Client-side scripting opens up the web to highly interactive sites, so as to allow the clients to carry out real-time interactions, including entry verification and simple calculation. Client side processing allows more complicated computation and user interface to be performed on the client machine rather than the networked server machines to avoid communication overheads. Server-side scripting and processing provide flexible approaches to data handling and system integration. Time-consuming computations are usually performed on server machines that are comparatively more powerful. The user only needs to submit the input data through the web browser and the server will return the results to the client browser. This would not only improve the computation performance but also allows the integration of intelligent and knowledge-based systems and databases into the web-based application systems.

Standard HTML pages are usually extended with more sophisticated components either attached to or embedded in HTML pages as application clients. Alternatively, client components are deployed as standalone systems. Standalone clients must be installed and configured before use. They must also be properly maintained and upgraded. Web browsers can be used to facilitate these activities. In contrast, web clients are first downloaded from the web servers and then automatically installed and executed by the web browsers on the client machines. While the standalone client needs only be installed once, the web client must be downloaded and executed on the client every time when the user connects to the web page if the client component has not yet been downloaded. If the client is "fat", then the downloading process may take a long time.

Web servers provide a number of ways of deploying server components, for example, CGI, Active Server Page (ASP). The standard HTML specification provides simple mechanisms for clients to call servers for certain actions.

In addition, application server components can be deployed on server machines other than the web server. For non-HTML/HTTP clients and servers, DCOM, RMI and CORBA are the three distributed computing specification models widely used at present for developing and deploying component objects on the server and client sides. Since they are at the specification level, DCOM and CORBA are supported in a wide variety of programming languages such as C++, Java, and Visual Basic.

From our initial experience, the combination of non-HTTP/HTML application servers and clients attached to or embedded in HTTP/HTML web browsers is a good choice in terms of both implementation and deployment.

4.4. CONNECTION OF REMOTE DATABASES TO WEB APPLICATIONS

Many product design and manufacture applications extensively involve database management. Although Object-Oriented Databases are perceived to be naturally more suitable for product design and manufacture applications (Adiga, 1993), relational databases are still most widely used for their well-established and indeed standardised SQL (Structured Query Language). There is a wide selection of commercial relational database management systems (RDBMS). For example, Microsoft SQL Server, ORACLE, and SYBASE are typical RDBMS with a client-server architecture. Microsoft Access, ProFox, etc. are also widely used in practice. Most of such popular databases are ODBC (Open Database Connectivity) compliant.

4.4.1.   ODBC and JDBC

ODBC is an open standard that provides a common set of Application Interface (API) calls to manipulate databases. Application developers can write applications that make ODBC calls. These applications will also be able to work with many databases, unlike those programs specifically written for a particular database using its native database APIs. Applications built with ODBC can take advantages of the majority of the functionality of the database. They are as fast as applications developed using native database drivers. In addition, they can be ported easily by switching the ODBC driver instead of re-coding the entire application.

Java Database Connectivity (JDBC) is a set of Java classes that provide an interface to SQL (Structured Query Language) database. It provides uniform access to a wide range of relational databases. It also provides a common base on which higher level tools and interfaces are built in (Chapters 23 and 24: Orfali and Harkey, 1998). If an application program is developed using the JDBC API, it will be able to send SQL statements to any relational database virtually. To use JDBC with a particular database management system (DBMS), a JDBC Driver is required to mediate between JDBC and the database. This allows programmers to write a single database interface, which enables DBMS-independent Java application development tools and products, and allows database connectivity vendors to provide a variety of different connectivity solutions. JDBC drivers are either direct – sitting on top of the DBMS's native APIs, or ODBC-bridged – sitting on top of the ODBC APIs.

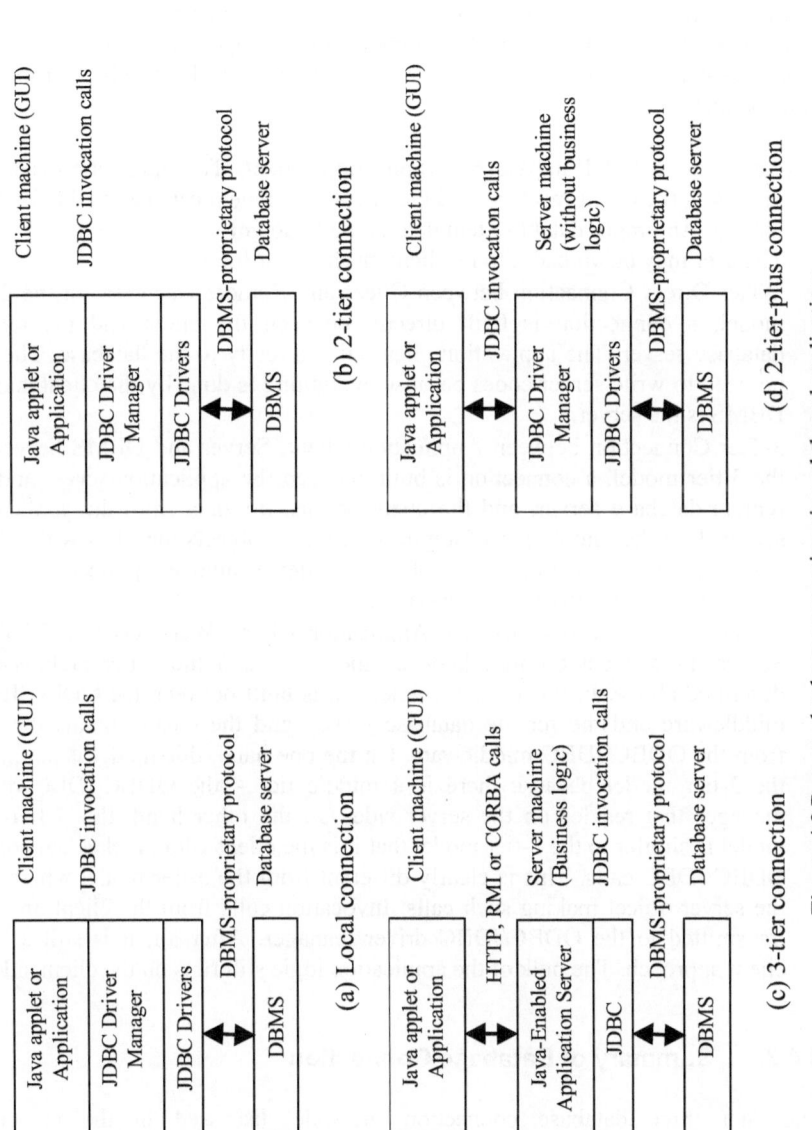

Figure 4.2 Remote database connections in web-based applications.

4.4.2.   Database Connections to Web Applications

The connection and access of remote databases are affected by many factors, such as the design and deployment of web application components. Figure 4.2 shows four possible ways for the user to connect and access the database from web applications:

- Access to Local Database by Application Client. The application database resides in the client machine and the client component can access the database locally. An empty database template is made downloadable from a web page. The user may download it to its client machine for future uses.
- 2-Tier Direct Connection between Client and Remote Database. In the 2-tier model, a connection is built directly between the client and the remote database server. The application client talks directly to the database. There is no need to write server codes because everything is done by the client and the DBMS is the server.
- 3-Tier Connection between Application Client, Server and DBMS Server. In the 3-tier model, a connection is built between the application server and the remote database server, and the client obtains the data from the application server. It is the middle tier of application server objects that deal with all the database access operations. Application clients invoke application server objects that in turn invoke the DBMS servers.
- 2-Tier-Plus Connection between Application Client, Web Server and DBMS Server. It is a compromise between the 3-tier and the 2-tier architectures described above. In this case, a connection is built between the ODBC/JDBC middleware and the remote database server, and the client obtains the data from the ODBC/JDBC middleware. On the one hand, this model is similar to the 3-tier model because there is a middle tier – the ODBC/JDBC driver manager that resides on the server side. On the other hand, this 2-tier-plus model is similar to the 2-tier model that it is the client who invokes the normal ODBC/JDBC calls. This is clearly different from the 3-tier model where it is the server object making such calls. Invocation calls from the client are then transmitted to the ODBC/JDBC driver manager. After all, it is still a "fat" client approach. The bulk of the application logic still runs on the client side.

4.4.3.   Summary of Database Connection

The first three database connection methods discussed in the preceding subsections have been used in our research projects. Based on our initial experiences, several insights have been gained. For example, the web-based "Design for Assembly" project (Chapter 8). One of the advantages of this approach is that the user has the full control of the database and its access. One notable disadvantage is that other clients are unable to share the database. Therefore, this model is suitable for those web applications where the users would like to maintain the database themselves.

The web-based "Failure Mode and Effect Analysis (FMEA)" project (Chapter 9) and the web-based "Design for X guidelines" project (Shi et al, 1999) use the direct connection between the application client and the remote database server. One notable issue encountered so far is that ODBC/JDBC drivers must be installed and properly configured on the client machine. Although this is straightforward, a user may not be aware of this or unwilling to do this before using the application or applet.

The 3-tier connection model is being investigated for the web-based synchronized "Morphological Charts" (Chapter 12) and the web-based synchronized "Design for X shell" project (Shi et al, 1999). In this model, the ODBC/JDBC driver is installed and configured on the machine where the server object resides. It is, therefore, not necessary for the client machines to be equipped with the ODBC/JDBC drivers. In addition, all clients share the same application server. Therefore, this sever can be used to maintain data consistency and integrity, so as to achieve synchronisation between different clients.

The 2-tier-plus model has also been investigated, using Active Server Pages and the group has decided not to investigate further at present. Obviously, this 2-tier-plus model overcomes some of the shortcomings of the 2-tier model and inherits some of the advantages of the 3-tier model.

To sum up, which method is most appropriate for a specific application depends on the nature of the application.

4.5.   SUMMARY

Researchers are increasingly interested in encapsulating their new techniques and methodologies for product design and manufacture in web-based prototype systems. This chapter has discussed a number of issues that must be dealt with in developing and implementing proof-of-the-concept web applications. The first issue is whether the web application under development should follow the 2-tiered or 3-tiered client/server architecture. Although the 3-tiered architecture is generally advocated for its performance and flexibility, extra efforts are required to develop and deploy the application server. In contrast, no application server is involved in the 2-tiered architecture with a "fat" client.

The second issue is which programming environment should be used to develop application clients and how should they be deployed. Web clients can be implemented in HTML pages supported by standard scripts, such as VBScript and/or JavaScript. HTML pages may be further extended with either ActiveX or Java components as embedded or attached objects. Visual Basic and Java are two simple-to-use programming languages for building ActiveX and Java components respectively. Web clients are usually deployed on the web server, downloaded, installed and executed on the client machine automatically during the access through the web browser. Alternatively, a client may be deployed as a standalone system.

The third issue is which programming environment should be used to develop application servers if they are involved, and how should they be deployed. CGI programs are predominantly the most popular choice for deploying application servers. CGI programs can be implemented in any programming languages virtually. Servlets are extensions to CGI made in the Java environment to implement and deploy application servers. ASPs are extensions to CGI made in the Microsoft ActiveX environment for developing and implementing application servers. Alternatively, application servers can be developed and deployed following standard distributed computing paradigms such as DCOM, RMI, and CORBA.

The fourth issue is how the application client interacts with its server(s). CGI, Servlets and ASPs all depend on HTTP through HTML web servers, as long as they are deployed following the convention. It is becoming increasingly easier to program DCOM, RMI or CORBA clients to call the server counterparts. There are situations where the application server needs to interact with its active clients. A typical application is the synchronous teamwork, meaning that the change made by one client is spontaneously reflected to other clients. It is difficult to achieve this through CGI programs and ASPs. It would be more straightforward to follow the DCOM, RMI and CORBA to deploy and treat the application clients as servers as well.

The last issue is concerned with the development and connection of remote databases that many web applications use in product design and manufacture. This issue is closely related to the first issue of the client/server architecture. In the 2-tiered architecture, the client connects directly to the remote database. In contrast, it is the server that connects to the remote database in the 3-tiered architecture.

In this chapter, discussions on the issues have been given as initial considerations when developing prototype proof-of-the-concept web applications in product design and manufacture. More factors such as security, ownership, access, etc. and their relationships must be considered when developing serious web applications to a commercial standard. In the final analysis, the experiences and skills accumulated over times must have important roles to play.

5

SYNCHRONIZATION OF WEB APPLICATIONS

As discussed in previous chapters, the Internet and web have been regarded as one of the most promising technologies that are expected to have profound impacts on product design and manufacture. Though significant progress has been made, the developments are still at the starting phase. Conventional web applications generally address two dimensions of distribution, namely geographical and specialisation. However, they do not sufficiently address the time dimension of distribution. This is mainly because the traditional approach of the Internet technology does not directly support the collaborative forms of information sharing (Bentley et. al., 1997). Accessing information on the web is unidirectional, asynchronous, and limited by a simple client/server model in which only predefined data are provided (Frivold et. al., 1995). Obviously, this kind of information flow is not suitable for supporting the geographically distributed working teams. Instead, bi-directional, synchronous web-based design tool is necessary for providing awareness inside and outside the group and discussing the concept in different geographical locations simultaneously (Tuikka, 1997).

Assume team members of the project are using the same web application individually for their own activities at different locations but at the same time. Individual clients of the same web application usually maintain their own workspaces locally on their client computers. Their decisions are first recorded in their own workspaces before being transformed into the central database, if any. This creates a number of problems, for example:

- The member users are limited to their own workspaces and therefore, it is difficult for them to reach a global optimum.
- The change made by one user cannot be reflected to other users instantly.
- Clients would not know the most up-to date progress made by others.
- The integrity and consistency cannot be maintained.
- It is impossible or extremely difficult to resolve disagreements and reach consensus decisions.

In order to address the above limitations of conventional web applications, this chapter will emphasis on the synchronisation of clients of the web application. This chapter will start with clarifying the requirements for synchronised web applications. Then, a number of the research issues in synchronised web applications will be addressed in the third section. Afterwards, conflict resolution techniques will be generally discussed. Finally, this chapter will report on how data synchronisation is implemented programmatically.

The research work reported in this chapter has already been carried out when the ProDefine system reported in Chapter 12 was under development. Therefore, these two chapters should be read together. Readers are suggested to acquire some brief understanding of the ProDefine first before investigating this chapter on web application synchronization.

5.1. COMPONENTS OF SYNCHRONISED WEB APPLICATIONS

To support the collaborative design over the Internet, it is essential to guarantee that all the project team members are working with the most up-to-date version of the design task. Moreover, there should be enough information exchange between different users so that every user will be notified of the progress of the other users (Mills, 1998). Therefore, data synchronisation is necessary for maintaining the integrity and consistency.

Let us consider the situation shown in Figure 5.1. Assume three clients are using the same web application for the same project. Among them, Client 1 is proposing a decision that actually changes the status of the project. If this web application is synchronized, this change must be reflected to the other two clients, namely Clients 2 and 3, instantly. Next, Client 2 proposes another decision based on the previous decision made by client 1. This time, the change is seen by Clients 1 and 3. This process repeats itself until the project is completed successfully.

Computationally, the above synchronization process is achieved through shared or synchronized workspace. In this book, a "synchronised" web application is a web application whose clients share the same workspace. Synchronisation is the real time replication of data on the design task to all the designers working on the same project. This guarantees that all the designers are working with the most up-to-date version of the design task. Any changes or modifications made to the workspace are reflected instantly in local clients' workspaces and visualised by the users. Further decisions can be built upon previous ones. Any conflicts and disagreements are resolved at appropriate times. This creates truly a virtual teamwork environment where members participate from different professional backgrounds, locations and times while integrity and consistency are maintained in their workspace.

In product design and manufacture, it is a common practice for a project team to hold meetings to carry out innovative design or critical design review. Virtual Teaming or tele-conferencing is widely used in major corporations for this type of meeting where members do not really come to the same physical meeting room. Assume that members of a project team are supported by a synchronized web application that can simply be video facilities or some more sophisticated decision support systems. The team follows an agenda of the meeting. The agenda is in turn split into topics or tasks and different members are responsible for different tasks. The process of this type of synchronised application on the World Wide Web is very similar to the meeting between project members located in the same room with similar equipment and software supports. The only difference is that members are distributed at different locations.

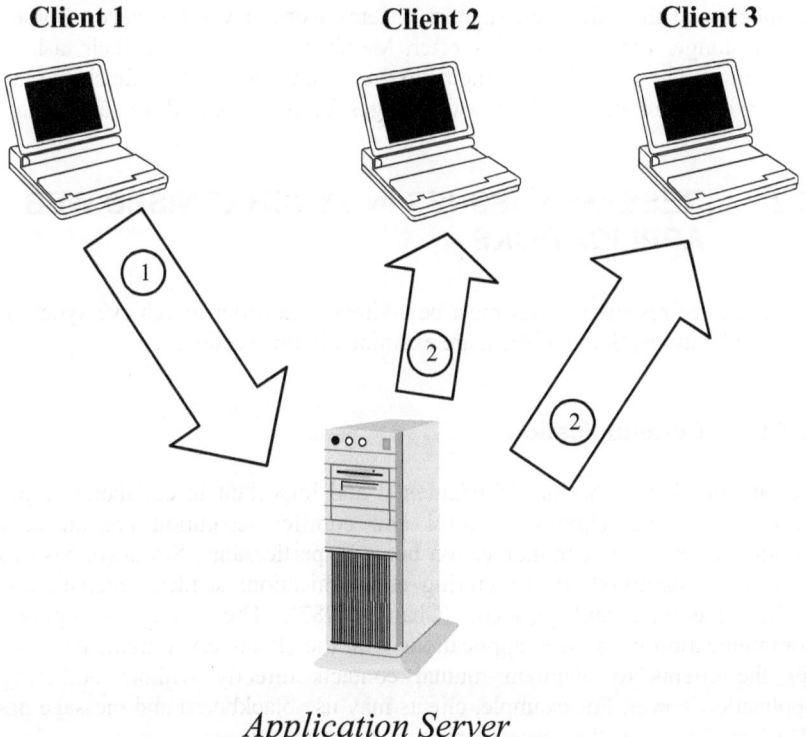

Client 1   **Client 2**   **Client 3**

Application Server

Figure 5.1 Synchronized web application.

At the meeting, the team follows the items in the agenda. Members make their own contributions (typically at the brainstorming sessions) and then the team reaches a consensus on an issue (if necessary by voting). For this kind of synchronised web application,

- Members must be (virtually) present at the meeting,
- All individual members must follow the same agenda and work on the same item at a time.
- Individual members (and their decision support systems) can only proceed at the same pace.

However, there are some situations where the above type of synchronised application is not suitable to apply. For example, it is not always suitable to require team members across the Atlantic and Pacific Oceans to be virtually present at the meeting because their office hours are totally different. Furthermore, individual members are usually charged with different tasks. It is not efficient to demand them to wait for other members to complete his or her task. This completely defeats the purpose of web application. Instead, they should work concurrently as far as possible to maximise the efficiency. In addition, individual members and their decision support systems work at varying paces. Some tasks require longer time and some shorter. Members seldom finish their tasks at the same time. They should start their next tasks only when they finish the tasks at hand. These limitations raise new challenges for synchronised web applications.

5.2.   RESEARCH ISSUES IN SYNCHRONISED WEB APPLICATIONS

A number of research issues must be addressed in order to achieve synchronised web applications. Some of them are examined in this section:

5.2.1.   Communication

Communication is the most fundamental and important in collaborative product development. Participation, control and conflict resolution are all achieved through an effective communication between participants. Seven (or Six) factors have been identified for structuring communication: sender, receiver, content, timing, medium, and protocol (Chang, 1987). There are two options for communication in the web applications with the client-server architecture. One is for the clients to maintain mutual contacts directly without bothering the application server. For example, clients may use blackboard and message passing (Decker, 1987). In the message-passing paradigm, messages may be selectively targeted to specific receiver(s), or at the request of a receiver, or broadcast to all participants, regardless of their interests. In blackboard paradigm, messages are routed by a blackboard metaphor with or without target receivers. Blackboard can also serve as a control mechanism (Englemore and Morgan, 1988). Alternatively, communication between clients, using blackboard or message passing, may be channelled through the shared application server. This latter approach would simplify the client design and implementation, so that clients may also be "thinner" thus quicker to download.

5.2.2. Role Assignment

Team members such as managers, designers, production planners, and supplier and customer representatives usually play different roles in a collaborative product development project. Role assignment is essential for dealing with the specialisation dimensions. This is usually sorted out during the stage of project planning.

In synchronised web applications, role assignment also defines different privileges of different members during the process of application. For instance, design manager is the only role that can use the whiteboard and customers can only observe the progress of the design work etc. This role assignment allows different users to have different rights in the workspace.

Combining with the role assignment, the user authentication system forms a basic access control mechanism in the workspace. Usually, if the collaborative system is in HTML, CGI or ASP bases, the web server can authenticate the designers by using username and password. The identified user will have a restricted power in the system and the username is used in the awareness information and the user tracking. This constitutes the basic security of the system.

5.2.3. Awareness of Presence

The awareness of the presence of team members in the project session is essential for collaboration. Awareness is the understanding of other group members or users' activities, which provides a context for your own activity (Dourish and Bellotti, 1992). It is a simple event service, which provides details of actions done within a workspace (Bentley and Applet, 1997). Awareness can be classified into high- level and low- level. High- level awareness only provides the actions of the other users while low level awareness provides the details of the actions (Dourish and Bellotti, 1992). Choosing among different types of awareness depends on the requirements of product design operations. The basic type of awareness may include a list of currently available participants. According to the details of the design work, the awareness can be an image of any user's screen that is updated several times per minute (Roseman and Greenberg, 1996). This method is suitable for design work dealing with 3-D model. Other techniques can be verbal description of actions performed by a user. An example of these may be 'User A changes the height of the product'. Additionally, some fine- grained awareness function might also improve the quality of collaboration. Users might need to know who is in awareness and to which type of information. Besides, users need to be aware of failures. Some special actions will be necessary if one of the collaborators loses connection to the system (Hall et. al., 1996). Moreover, different (participant) roles may need different awareness requirements.

5.2.4.   History Management

Product development usually involves long time interval. Designers may need to remember the reasons underlying the design decisions made during the design procedure. This rationale information is useful in modifying the previous work with similar challenges or goals (Klein, 1995). Therefore, product development systems will need to provide some backtracking facilities for the designers to retrieve the design history and principles. This allows the designers to have more understanding on how the product design is evolved. Existing rationale information can be in the form of graphs, graphics or text descriptions (Lee and Lai, 1991). This information usually represents the reasons or principles for decision- making. In supporting the retrieval of the rationale information, designers may need to drop down their design intent or reasons behind each decision made during the product development process. However, it is necessary to reduce the workload of the designers involved in the rationale descriptions and to reduce the tendency to waste time on irrelevant or unimportant issues. In addition, decisions are made by some participants at the absence of others who may disagree with their decisions. A historical record would assist late participants to assess the impacts of early decisions and make appropriate contributions.

5.2.5.   Participation

There have appeared a number of strategies for facilitating participation of human members of a project team. Many of them have been attempted for computer implementation. For example, the concept of agenda has been used as a prescriptive strategy to involve participants in certain order. The brainstorming and synectics metaphors have not been investigated widely as a computational strategy for participation although they have been widely used as a general innovation technique. The concept of contract-net has been investigated and attempted as a computational participation strategy (Parunak, 1988). The underlying principle of contract-net is similar to the tendering process. A goal is advertised and all participants bid to achieve the goal. A concept similar to contract-net is negotiation. Negotiation is a process in which the participants iteratively exchange proposals and proposal justifications until an agreement is reached (Sycara, 1989). The process involves three main tasks: generation of a proposal, generation of a counterproposal based on feedback from dissenting participants, and recommendation of justifications and supporting evidence.

5.2.6. Conflict Resolution

Participant systems address problems from their own perspectives. Conflicts are naturally unavoidable at some stages. The integrity and consistency must be maintained between participants as collaborative design progresses although local and temporary inconsistency is both allowable and necessary to enable overall optimisation.

Collaborative product development involves exploration and backtracking in order to propose solution plans and refine them, and to diagnose and remedy failed solutions. A control mechanism is usually needed to maintain a high degree of convergence towards an overall good solution while encouraging and accommodating divergent contributions. Tong (1987) proposed three types of principles in addressing this type of control. First, *the least commitment principle* implies that a final decision should not be made arbitrarily or prematurely but postponed until there is enough information to do so. Second, *the early commitment principle*: This implies that decisions should be made as early as possible and any conflicts occurring later will be addressed through backtracking or iterations. Finally, *the opportunistic commitment principle*: This provides a promised strategy for controlling the decision-making. Decisions are made as required. It moves attention opportunistically from one participant to another.

5.3. STRATEGIES FOR CONFLICT RESOLUTION

Ideally, multiple users (clients) in a synchronized web application should collaborate to converge towards a common objective. In reality, their decisions contradict, as well as complement, with each other. This is inevitable because of their divergent disciplines and expertise, and sometimes with different objectives. Conflict resolution, computationally or manually, is very complicated. Nevertheless, certain actions may be taken within the web application to adopt appropriate strategies.

According to Klein, Lu and Baskin (1990), there are two categories of conflict situations. They are competitive conflict situations and co-operative conflict situations. In co-operative situations, different parties are co-operating with a common goal to achieve a globally optimal solution. In collaborative product development applications, share a common goal for the different expert domains does improve the product design. This type of resolution involves techniques for finding the solutions with mutual benefits.

Very often, a web application does not have to provide any specific mechanisms for cooperative or proactive conflict resolution. Instead, conflicts are usually avoided before solutions are generated. It is achieved by fully specifying the conditions or constraints upon which solutions are worked out. For example, the cost engineer specifies a constraint on the cost range and the reliability engineer specifies a reliability requirement. The two constraints or requirements are used as the conditions for the "Product Definition" web application to generate conceptual solutions. The resulting concepts should meet both requirements and therefore, they are no longer contradictory to each other.

Conversely, in competitive situations, different parties work on their own interests or benefits only. They agree in one way and disagree in another, with each other's individual contributions. Therefore, there can be conflicts of interests due to the interactions of the multiple diverse experts. For example, the marketing experts will put the focus on the popularity and the functions of the developed product while the design experts put their focus on the feasibility of the system. The mechanical designers may have different viewpoints or ideas from the electrical designers. Since they are working in a virtual space rather than the same place, this brings the necessity to provide a number of means to prevent or resolve the conflicts.

In order to deal with competitive conflicts, the following four facilities are commonly used to support several strategies:

- Shared common workspace for synchronised maintenance of data and information integrity.
- Locking Mechanism for maintaining data integrity.
- Decision Fusion Explorer for resolving competitive alternatives in numeric terms.
- Vote Explorer for resolving competitive alternatives in qualitative terms.

Chapter 13 presents the ProDefine web application that uses the above approaches in detail. The remaining discussions in this section are based on this specific web application for product definition.

5.3.1.   Synchronisation Using Shared Workspace

Before we discuss how to deal with conflict resolution, it is necessary to address the issues of maintaining the data integrity between different users in order to avoid conflicts. While product development involves multiple functional perspectives, maintaining a data consistency control mechanism can be a significant challenge on system development (Bentley et. al., 1997; Klein, 1995). Fail to provide consistent information may have a direct impact on the product cost, quality and development time.

The workspace is a set of data objects. Each object represents certain portion of the decision space of the product development project. For example, the Customer Requirement Explorer maintains its own workspace through a data object such as a Treeview ActiveX control or an equivalent relational data table or resultset from an SQL query. Likewise, the Concept Generation Explorer maintains its workspace using the morphological charts or ActiveX grid control. The ProDefine application server also maintains its workspace using the same or similar data objects.

Basically, each client maintains its own local workspace. Changes are first made to the local workspace and then sent to the ProDefine application server.

The ProDefine application server updates its own workspace accordingly in response to the client's request. The ProDefine application server will then broadcast a message to all the current clients, requesting them to update their local workspaces. As a result, the ProDefine clients update their local workspaces accordingly. The process repeats itself with further changes in the status of the workspace.

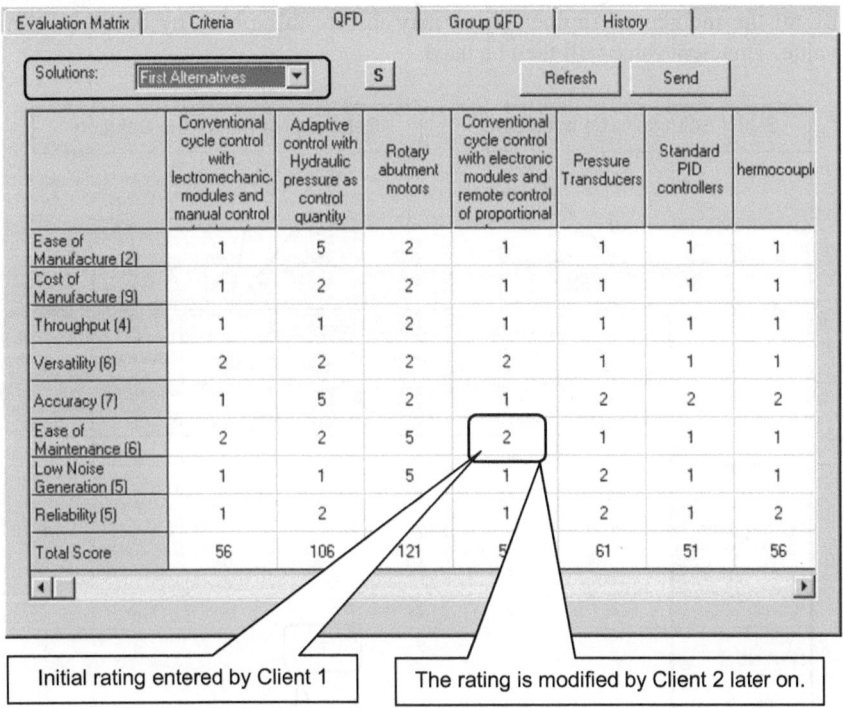

Figure 5.2 The Group share a single workspace.

5.3.2. Decision Fusion Explorer

Very often, individual contributions made by team members are numerically represented. Take the example of concept evaluation, each member is invited to evaluate the alternative concepts independently. Thus, the evaluation results are surely different, i.e. conflicts exist. ProDefine provides some methods for calculating numeric consensus:

- The group shares a single synchronised worksheet.
- The members maintain their own worksheets to work out their individual overall results, which are then combined to form the overall group result.

- The members share the single synchronised worksheet and their intermediate individual results are combined first to form the group intermediate results which are finally combined together to form the overall group result.

The above three methods are shown in Figures 5.2, 5.3, and 5.4 respectively. The first method can be implemented by using the Group QFD tab utility in Figure 5.2. A client may assign an initial rating in the matrix. This initial rating is shared by all the members. Another client may update this rating by entering another value. This new value will then be used.

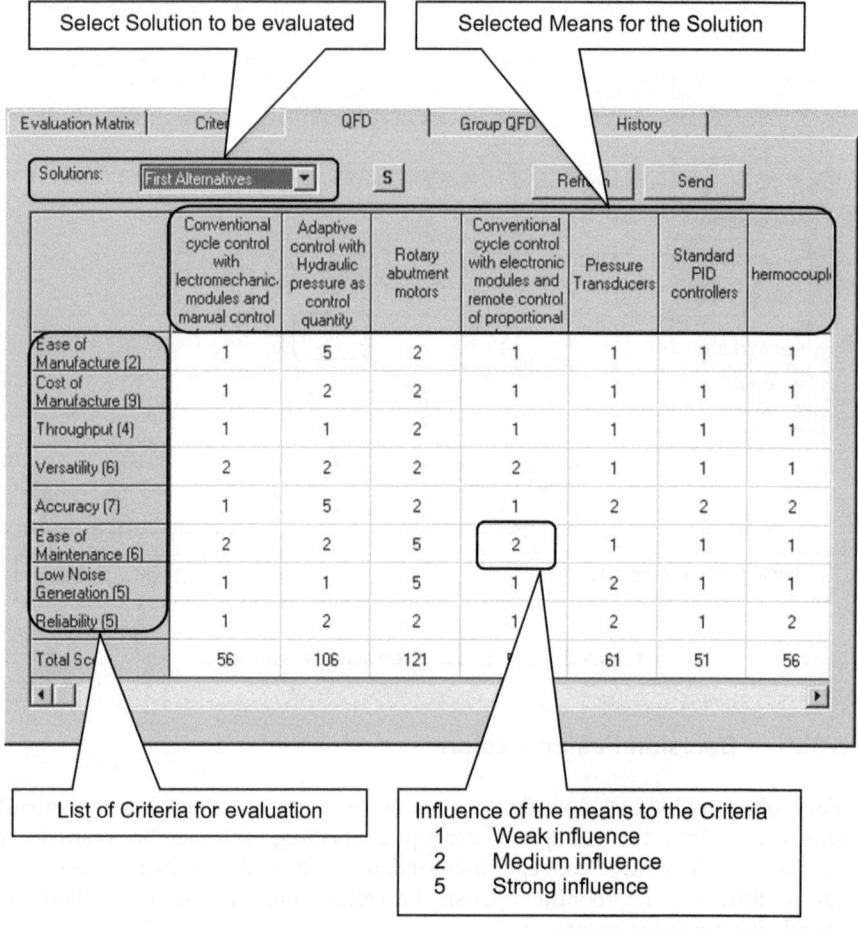

Figure 5.3 Members produce their individual results first and then combined to form the overall group result.

The second method is to use the average value of all the individual evaluation results as the consensus value. The method allows designers to maintain their own evaluation workspace. Each designer can evaluate the solutions and give ratings independently. The results of the evaluation are then sent to the application server. The application server will collect all the personal evaluation ratings and calculate the group average. The group average is then distributed to all the clients and considered as the group result.

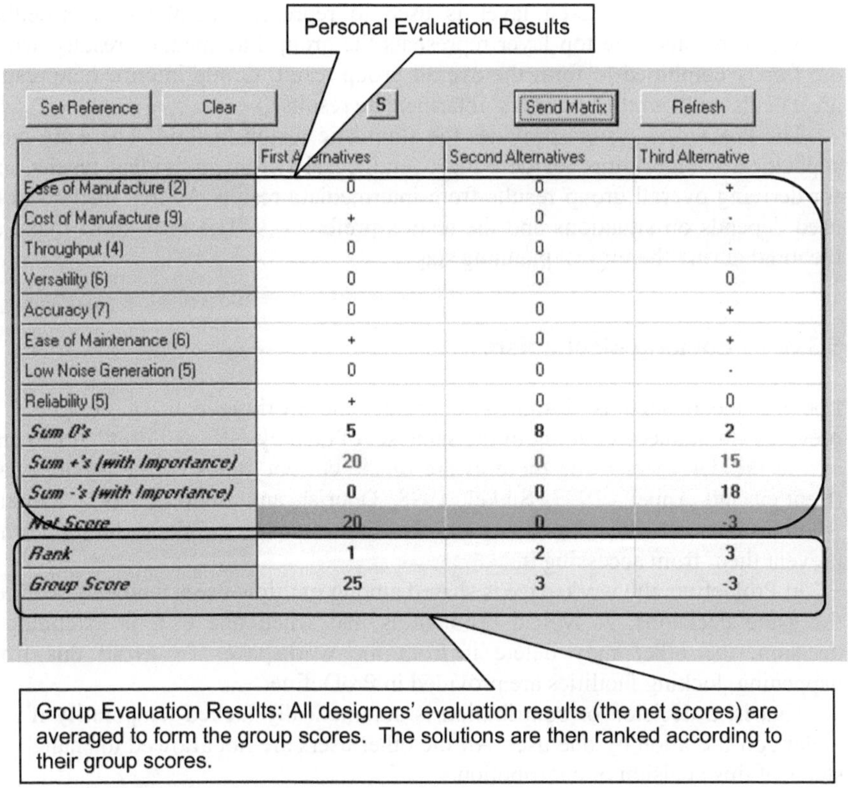

Personal Evaluation Results

	First Alternatives	Second Alternatives	Third Alternative
Ease of Manufacture (2)	0	0	+
Cost of Manufacture (9)	+	0	-
Throughput (4)	0	0	-
Versatility (6)	0	0	0
Accuracy (7)	0	0	+
Ease of Maintenance (6)	+	0	+
Low Noise Generation (5)	0	0	-
Reliability (5)	+	0	0
Sum 0's	5	8	2
Sum +'s (with Importance)	20	0	15
Sum -'s (with Importance)	0	0	18
Net Score	20	0	-3
Rank	1	2	3
Group Score	25	3	-3

Buttons: Set Reference | Clear | S | Send Matrix | Refresh

Group Evaluation Results: All designers' evaluation results (the net scores) are averaged to form the group scores. The solutions are then ranked according to their group scores.

Figure 5.4 Intermediate results from individual members are combined to form intermediate group results which are then combined to form overall group result.

There are a number of advantages in using this implementation. First, this method eliminates the conflicts during the evaluation stage because designers are working individually. Second, this method is fast and simple since there is only a little or no discussion at all during evaluation. Third, only the average results of the group will be visible to the designers. The ratings of individual designer are kept secret. Therefore, designers' decision will not be affected by others, and they are free to evaluate according to their own knowledge and experience.

However, certain disadvantages are expected in using this method. First, no discussion is involved during evaluation. Usually, discussions will lead to more understanding of the product. New ideas may evolve during discussions. This method eliminates the chance of getting improvements to the new design. Second, designers cannot review others designers' evaluation data. This indicates that designers do not know how the group result is being calculated. Users are not allowed to backtrack the results even when designers doubt the final result.

In the third approach, the members share the single synchronised worksheet with multiple layers. Each layer is used to represent members' intermediate individual results. The top layer represents the group intermediate results, which are finally combined to form the overall group result. Group intermediate results are calculated from the members' intermediate results.

The ProDefine system provides the simplistic fusion method. There are more sophisticated algorithms for deriving overall results from individual results, and for deriving overall group results from intermediate results. Which method to be used depends on situations and the team's preferences. The team must make up the mind during the project planning stage.

5.3.3.   Locking Mechanism

Locking mechanism is designed to protect the working data of designers. It prevents simultaneous editing of the same set of data by different users. It can also ensure the designers to be more aware of the current status of the product data (Bentley and Appelt, 1997, Sikkel, 1998, Dourish and Bellotti, 1992). Usually when the data is locked, no other user can edit the same set of data. But it does not prevent them from accessing it.

In ProDefine, the workspace is shared among multiple users who may work on the same decisions. A typical problem is that when one user is extending a decision, the other may delete it from the workspace. To avoid this from happening, locking facilities are provided in ProDefine.

That is, a decision or contribution is automatically locked temporarily if it is under consideration by one user. All the other users are not allowed to change the status of this decision or contribution.

Permanent locking may be used to prevent any actions or events that might change the status of the decision concerned. One possible application of permanent locking is to lock the design decisions that have been formally reviewed by the project team. Thus, no changes can be made without the team consensus.

There are some general issues concerned when designing the locking mechanism. First, the levels of locking should be designed carefully. For instance, in document writing system, it is necessary to decide whether the locking mechanism works on document-based, paragraph-based or sentence-based. Besides, it is common for the designers to analyse the product in the table format. A question here is should one cell or the whole table be locked if it is under editing. It depends on the nature of the design system. Second, the locking mechanism can be triggered automatically or manually. In some systems, the designers' working context is locked automatically when the edit cursor is put on the system. This may be too 'sensitive' because some designers may use the edit cursor to point out some important figures to their colleagues. Alternatively, the designers need to do some extra tasks in order to trigger the lock event. Some typical way to achieve this is to have a locking button that is responsible to lock the working context.

5.3.4. Vote Explorer

ProDefine provides a Vote Explorer or voting mechanism for resolving differences of a decision among the team members. This voting mechanism is only activated if the team chooses the strategy that all or certain types of individual contributions must be voted before they can become the team decisions. The Vote Explorer would not work if the team decides to adopt the "first- come- first- accepted" strategy. This agreement must be accepted or determined at the time of project planning.

When a user makes an individual contribution to the workspace, a record is made in the Vote Explorer. The contribution is marked as temporary decision in the workspace, pending the vote by the team members.

Individual team members may use the Vote Explorer to list all the individual contributions and choose some or all items to vote.

The voting is straightforward. If the member is in favour of the contribution, just click the "For" button to cast the vote. If the member is against the contribution, just click the "Against" button. If the member considers that the contribution does not concern him or her, then a "Don't care" button should be clicked. If the contribution receives supports from all the members concerned, it becomes or is accepted as the team's decision and its status changes from "Temporary" to "Permanent" in the workspace. If the contribution is refuted by all the members, then it will be removed completely from the workspace. However, these two situations of full support or full disagreement are unlikely in practice. Therefore, certain criterion is needed to define "acceptance" and "rejection". Some possible and simple criteria are "majority voting" and "50% votes". Again, elements of criteria must be determined by the whole team during the project planning stage.

Web-based collaborative design system usually involves group decision support. Voting is being used in most systems to indicate group consensus. The product development system should allow group members to vote electronically with the selection of "agree", "disagree", or "abstain" on the others' work (Hanneghan, Merabti and Colquhoun, 1996). The system is then responsible for collecting the voting results and displaying the outcome. This information together with the reasons behind each decision can be part of the rationale information. It should be retained within the repository for further review.

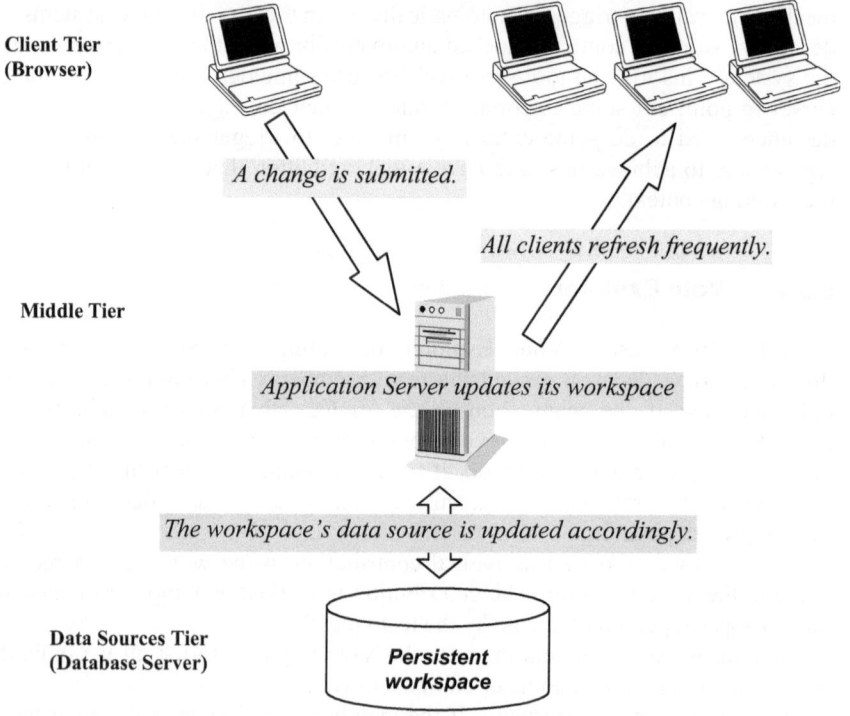

Figure 5.5 Frequent refresh of the central workspace by the clients.

5.4.   SYNCHRONISATION IMPLEMENTATION

Two methods have been commonly used for implementing the synchronization of web application. The first one is to frequently refresh the client browser so that the most up-to-date information is downloaded from the application server. The second one is for the client to notify the server and update its central workspace. The server in turn updates the local workspaces of individual clients.

5.4.1.   Client Refreshing from Server-to-Client One-way Communication

A client submits a request to update the central workspace maintained by the application server component (and accordingly the data source as well), as shown in Figure 5.5. If other clients refresh their web browsers with the same application client, the server will send the most up-to-date information stored in its central workspace to them. If such refreshing takes place at a very high frequency (say a second), then the users at different locations are under the same impression that they are working simultaneously. This type of synchronization has been widely used in implementing chatting facilities and electronic whiteboards.

The implementation of this type of synchronization is relatively straightforward. For example, an invisible Java applet is embedded in a client web page. This Java applet periodically refreshes the web page at a specified time interval. It is very simple or very "thin" and therefore fast to download.

With this type of synchronization, there is no need to maintain a local workspace by the client. Hence, the client is relatively "thin" and therefore very quick to download. Since all the operations are performed by the application server, only the client is able to initiate communications with the application server. The application server has no way to communicate with the clients unless the clients start the communication. One metaphor of this situation is the server does not have the phone numbers of the clients but the clients have the phone number of the application server.

Strictly speaking, it is not synchronized! Anything could happen during the short refreshing interval. However, this approach seems to work well in many applications.

Figure 5.6 shows the code example of refreshing implemented in ProDefine. This is one-way communication where the ProDefine clients request services from the application server. In ProDefine, a class object, which is a type of ActiveX code components, is created in the application server. It consists of a list of business logic and data request services. The application client then creates an instance of the server class object by locating the IP address of the server. By this time, the application client is able to request the services from the application server by referencing the instance of server class object.

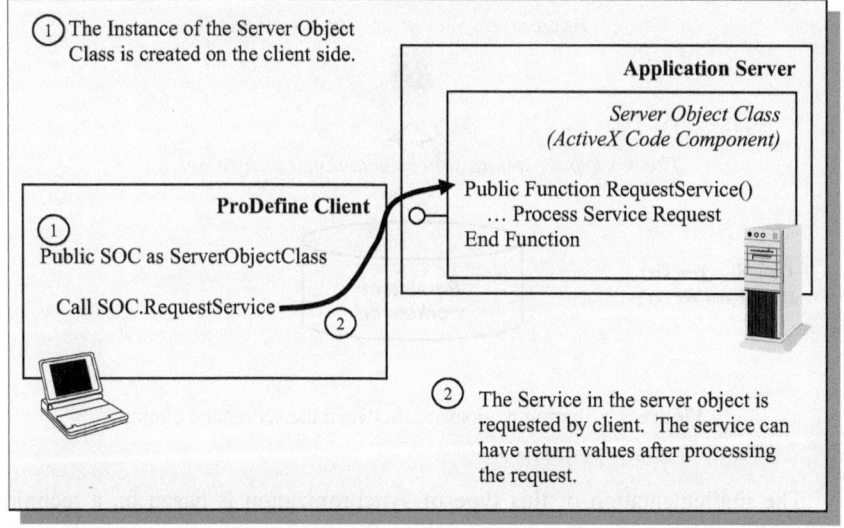

Figure 5.6 One-way Client to Server Communication.

5.4.2.   Client/Server Two-way Communication

Another approach of synchronizing a web application is to achieve two-way communication between the application server and multiple clients. In order to achieve synchronisation, the application server must be able to notify the clients for changes in real time actively.

Initially, a client proposes and submits a change to the application server to update its central workspace and the corresponding data source. Once this change is detected, the application server notifies its online clients to update their local workspaces accordingly. Basically, the clients' local workspaces are the clones of the server's central workspace. Therefore, all the users using the same client can see the change simultaneously.

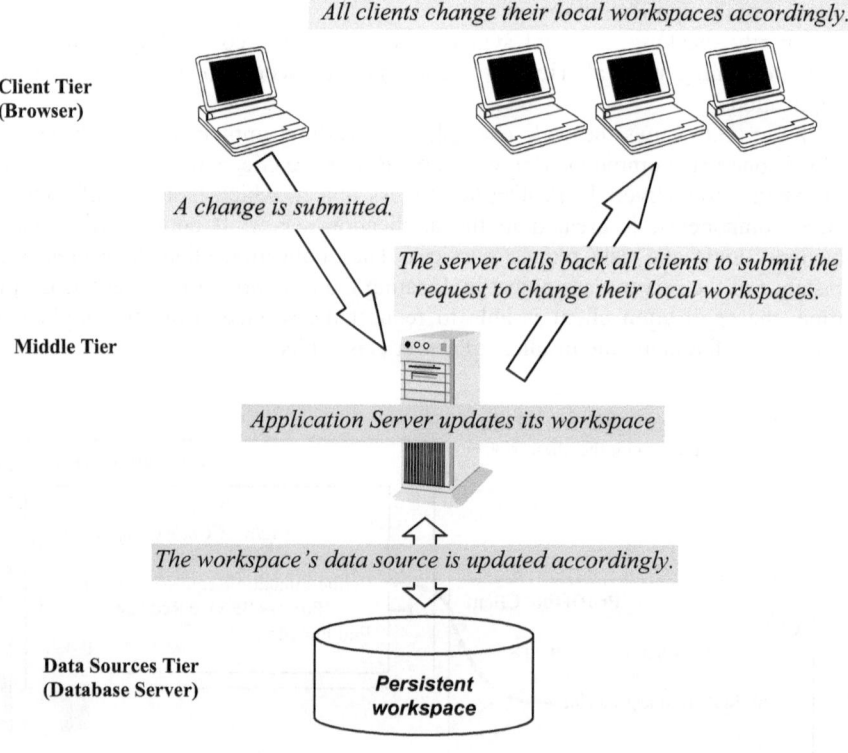

Figure 5.7 Sharing workspaces between the server and clients.

The implementation of this type of synchronization is based on a technique called 'Callback'. This technique enables the application server to initiate data communication with the clients. In 'Callback', the identity of the application server and the clients is interchanged. The application server can become the client and the client can become the server. That is, the application server requests for services from the client. More specifically, the application server requests the clients to change the data or values that have been modified by a client.

Figure 5.7 shows the code example of how the server initiates notification to the client. A class object is created on the clients. The object consists of many data modification services, from which the application server can request data changes. The application server then made a reference to that class. Whenever the application server is notified for any data modification, a data change request is sent to the client according to the client's object reference. Combining with the one-way client to server communications, two-way communication can be achieved.

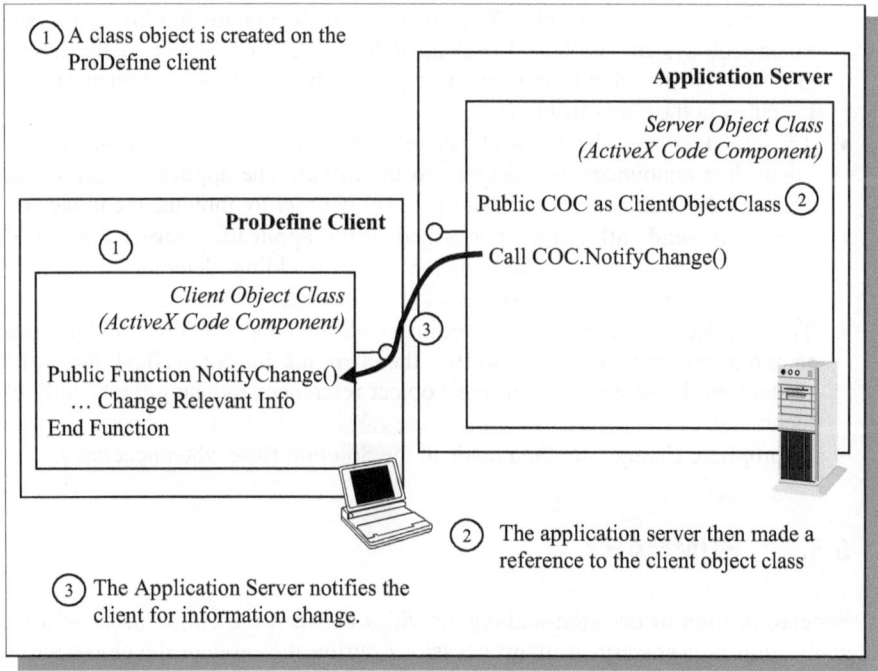

Figure 5.8 Server to Client Communications.

In order to truly achieve data synchronisation, the application server must be able to notify all the clients for data modification. Therefore, "callbacks" to multiple clients are necessary. One way for the application server to handle multiple clients is to keep all the references to the clients' class object, as each client has its own connection object. As a result, the application server notifies the clients one by one by tracking their object references.

In this approach of synchronizing a web application, clients maintain their own local workspaces and also associated operations for manipulating the workspaces. Therefore, the clients are relatively "fat" and thus the downloading process takes much longer time. In addition, the security setting of the clients is much more complicated within the web browser because the client becomes the server during the "callback" process. Our research group has spent a lot of time and efforts in resolving this issue. However, due to the inherent limitation of the technology, it remains unsolved at the moment.

To sum up, the general steps in the synchronized web application are as follows:

- The client first requests the homepage from the web server. In the case of the ProDefine system, the ActiveX document, which contains the Product System client-side system, is then downloaded from the Web Server to the Internet Browser in the client- tier upon request. The ActiveX document is then executed in the client machine.
- The client creates the class object reference of the application server. The client then announces its existence to the server. The application server then creates a class object reference of the client in order to implement callbacks.
- The client sends all actions performed to the application server through the server class reference. The actions may include adding, deleting or modifying of objectives, defining solutions, voting or locking.
- The Application Server then co-ordinates all the clients' actions and performs required calculations. It then notifies the client machines to refresh the current content on the screen through their object references by using the "CallBack" technique.
- Appropriate changes are then made to the Solution Base when necessary.

5.5.   SUMMARY

Synchronization of decision-making activities between the online users of a web application is necessary in many occasions during the product development and realization process. Although web applications support multiple users, who are geographically dispersed, special considerations are required when the synchronization facilities are introduced.

This chapter has discussed the requirements of synchronised web applications. A number of research issues for synchronous web applications are pointed out. They are communication, role assignment, awareness of presence, history management, participation and conflict resolution. Various conflict resolution strategies are addressed. The shared workspace, locking mechanism, vote explorer and decision fusion explorer are some of the facilities that have been widely used for this purpose.

The discussions in this chapter are introductory and elementary. Much work remains to be done if a synchronized web application is truly applied in product design and manufacture. One of the challenging issues is the granularity of synchronization, which is at which level of decision to be synchronized. We could synchronize events as detailed as mouse movements or mouse clicks. This would involve heavy interactions between the clients and server; thus, consume high overheads resources on communications. Alternatively, we could synchronize decisions, such as the selection of components or parts used in a product design. In this case, users are limited in participating in collaborative work. Finally, it is hoped that this chapter will stimulate further research interests.

6

BUSINESS MODELS OF
DIGITAL MANUFACTURING PORTALS

The Internet and the web are playing increasingly important roles in business, leading to the emergence of digital enterprises. The industry of digital enterprises has developed drastically in the last decade or so, and there have appeared a number of business models. However, these business models have not been scrutinized at the similar pace. Despite of some investigations, there is no formal definition of a business model. Furthermore, consultants and practitioners have often resorted to using the term "business model" to describe a unique aspect of a particular Internet business venture. In this respect, these attempts are neither complete nor robust. Questions remain to be addressed, for example,

- Can we apply the existing theories or frameworks to analyse the business models of digital enterprises?
- Are the business models of digital enterprises really different from those of traditional enterprises?
- What kind of new dimensions should be incorporated into the business models of digital enterprises and what should be the scope?
- How are the business models of digital enterprises related to those of the traditional enterprises that extensively employ the digital approach?
- Are existing business models of digital enterprises adequate to serve the industry, and/or do we need new models to cater the new developments?

This chapter is NOT aimed at answering the above- mentioned questions. Instead, the chapter aims to provide some basic understanding of the business models and their importance to digital enterprises, and review some of the efforts made in the literature and of course in practice. In a sense, the aim of this chapter is to raise further questions instead of answering them.

In order to better understand the business of Internet/web-based digital enterprises, it is useful to revisit the business models of radio and television broadcasting in the past century. The broadcaster is part of a complex network of distributors, content creators, advertisers (and their agencies), and listeners or viewers. Interactive television (TV) extends the scope even further.

Digital enterprises are able to accomplish all those elements provided by the traditional TV and radio broadcasting. In addition, they have new characteristics. For example, the Internet and web technologies involved are much more sophisticated. Complicated business and engineering activities that require sophisticated computation can be carried out over the Internet and web. Although traditional enterprises resort to Radio and TV for some business activities such as advertisements, they hardly incorporate Radio and TV into their business operations. In contrast, traditional enterprises are able to take advantages of the digital enterprise approach to maximise their competitiveness, increase their operational efficiency, reduce the cost and improve the quality. Consequently, the complexity of business models for digital enterprises will increase dramatically.

Back to the basic now, a business model is the method of doing business through that a company can sustain itself by generating revenue in the value chain. A firm can build and use its resources to offer better values to its customers than its competitors can. Therefore, it can earn more money than others. A business model details how a firm makes money   and how it plans to do so in the long-term, how a firm offers value to its customers, which segment of customers is targeted to offer the value, what scope of products/services it offers, what its sources of revenue are, how it prices the value that it offers its customers, in what form to offer the values, what capabilities these activities rest on, what the firm must do to sustain any advantage that it has, and how well it can execute (implement) these elements of the business model.

This chapter selects some of the facets mentioned above to examine the business models of digital enterprises. The discussion will focus on the following aspects respectively:

- Business players.
- The nature of digital enterprise.
- Functionality of web application/contents.
- Revenue generation method.
- Value generation method.
- Access control method.
- Visitor attraction method.

6.1.   BUSINESS PLAYERS

Manufacturing portals involve a variety of business players with different roles. These players gain their benefits from the concept of manufacturing portals, either by creating business opportunities, improving cost-effectiveness of the existing services, improving operational efficiency, providing better customer service and thus customer satisfaction, etc.

Some of business players who are somewhat directly involved are briefly explained as follows:

- Manufacturing portal operators are responsible for operating the manufacturing portal server, as well as the subscription and deployment of the web applications and management of users and their accounts in the capacity of service aggregators.
- Web application providers are responsible for providing services, ranging from simple contents to sophisticated interactive decision support systems.
- Technology solution developers/providers include suppliers of both hardware and software products used in the manufacturing portal. Hardware suppliers include manufacturers of fibre and equipment, intelligent manufacturing equipment and device, etc. Software developers provide portal technologies matching and bridging both the hardware equipment and user needs.
- Connectivity providers are responsible for providing infrastructure of the manufacturing portals, for example linking the enterprise's network to the Internet and therefore to the networks of business partners.
- All parties involved in the lifecycle of product development and realization process are potentially users of the manufacturing portals, depending on their roles and responsibilities.

The network of business players involved in manufacturing portals is potentially creating a new industry of engineering web application provision. However, it is in its very early infancy.

6.2.   NATURE OF DIGITAL ENTERPRISES

Business models vary with the nature of the enterprise. Digital enterprises are no exception at all and there are two categories of them:

- Category I: This category includes Web Applications Providers (WAPs). These enterprises conduct their businesses mainly by providing contents (which are the relatively simple form of web applications) and more sophisticated web applications. Strictly speaking, providers of the associated software and hardware technologies such as enterprise portal technology do not fall into this category.

- Category II: If traditional businesses such as manufacturers and software developers extensively employ web applications across the entire business process, then they become digital enterprises. These enterprises do not merely set up web sites, displaying information on the products that they sell in the physical world, but also conduct commercial transactions with their business partners and buyers over the Internet. Therefore, providers of the associated software and hardware technologies such as enterprise portal technology fall into this second category if they satisfy the above criterion. Similarly, a manufacturer becomes a digital manufacturing enterprise or simply digital manufacturing or e-manufacturing enterprise.

Vast majority of digital enterprises fall into Category I (DE I) although some of them are spun off from traditional enterprises. Naturally, most of the business models discussed in the literature belong to this category of digital enterprises. Table 6.1 lists some of the typical business models. More discussions are available from other sources, for example:

- http://ecommerce.ncsu.edu/topics/models/models.html
- http://www.frufalfun.com/5webincomemodels.html
- http://www.hiperworld.com/business-model.html
- http://www.insurance.about.com/industry/insurrance/library/weekly/aa042000a.htm
- http://www.workz.com/content/1148.asp

Business models for Category II digital enterprises (DE II) are much more complicated than those of Category I. One of the main reasons is that digital enterprises of DE II type usually have their own traditional business models that cannot be changed overnight or frequently. Also, they may not be consistent or flexible enough to accommodate the business model of DE I. Manufacturing companies are good examples. Therefore, the key issue here is to evolve and integrate the existing traditional business model with the emerging DE I business models to form appropriate DE II business models.

The business model for digital manufacturing enterprises is complicated by the fact that the enterprise portal may be composed of several sub-portals such as Procurement Portal, Sales Portal, Design Portal, Manufacturing Portal, Project Portal, and Customer Support Portal, etc. These sub-portals may have their own business models that differ from one another.

For example, the infomediary model should be the basic element of the enterprise web site. It is responsible for providing useful information to any party who is related to the business in someway. The e-shop model should be retained to provide the marketing function over the Internet. The e-procurement or e-purchase model should be retained to provide the purchasing function over the Internet. The brokerage model is used to facilitate online negotiation among different players, etc.

Some of the business models involve third parties to be responsible for business and technical information processing. E-marketplace is such an example. On the one hand, it is responsible for identifying the best overall solution from the centralized database that all the members share and contribute. On the other hand, much of the commercially sensitive data such as order patterns, technical specifications, etc. of a member enterprise is left at the central database, manipulated by a third party. If the information in the central database were being released to a competitor, the consequence would be disastrous. Therefore, caution should be taken with this type of business model regarding the confidentiality and security of commercially sensitive business and technical information.

Table 6.1 Timmers' eleven business models of electronic markets

Model	Description	Examples
E-Shop	Web marketing of a company or shop	http://www.fleurop.com/ http://www.travelocity.com/
E-Procurement	Electronic tendering and procurement of goods and services	Japan Airlines at Fehler!Textmarke nicht definitert
E-Auction	Electronic implementation of the bidding mechanism	http://www.fastparts.com/
E-Mall	A collection of e-shops, usually enhanced by a common umbrella	www.emb.com www.industry.com
Third party e-marketplace	Marketing by a third party on the web	www.tradezone.com
Trust Services		
Information Brokerage		
Value Chain Service Provider	Specialized function for the value chain such as electronic payment or logistics	www.ups.com
Virtual Community	Members with common interests share the information on the web	http://apparelex.com/bbs/index.htm
Collaboration Platform		
Value Chain Integrator	Integrating multiple steps of information flow in the value chain	PartnerNet MarshallNet

Source: Timmers (1998)

Table 6.2 Nine types of business models on the web

Model	Description	Examples
Brokerage: Bring the sellers and buyers together and facilitate transactions		
Buy/Sell Fulfillment		
Market Exchange		
Business Trading Community		
Buyer Aggregator		
Distributor		
Virtual Mall		
Metamediary		
Auction Broker		
Reverse Auction		
Classified		
Search Agent		
Advertising: Electronic extension of the traditional media broadcasting model		
Generalized Portal		
Personalized Portal		
Specialized Portal		
Attention/Incentive Marketing		
Free Model		
Bargain Discount		
Infomediary: Collecting and selling information about consumers and their buying habits etc. to other businesses		
Recommender System		
Registration Model		
Merchant Model: Electronic extension of the merchandizing model		
Virtual Merchant		
Catalog Merchant		
Surf and turf		
Bit Vendor		
Manufacturer Model: Websites allow manufacturers to reach their customers and suppliers directly on the web and therefore to compress the distribution channel.		
Affiliate Model: Providing purchasing opportunities wherever users may be surfing in affiliated partner sites.		
Community Model: Websites for users with common interests to share information.		
Voluntary Contribution Model		
Knowledge Network		
Subscription Model: Users pay for their access to the site.		
Utility Model: Metered usage or pay as you go approach		

Source: http://ecommerce.ncsu.edu/topics/models/models.html

Table 6.3 Five business models of e-commerce at Frugalfun.com

Model	Description
Vanity	Websites as outlets of self expression to share hobby, promote a cause, etc. with little intention of deriving revenue
Billboard	Websites designed to derive economic benefit through indirect means from either referred sales, reduced costs or both
Advertising	Websites for advertising products, services, etc
Subscriptions	Websites nurtured by publishers and accessed by the paid subscribers
Storefront websites	Websites offering products or services for sale

Source: http://www.Frugalfun.com/5webincomemodels.html

Table 6.4 Insurance e-commerce business models

Model	Examples
Brochure site	Websites of companies or agencies simply serving as business cards on the web.
Customer service site	Limited but direct interactions with customers.
Real time site	Websites allow the customer to access rate and account information in real time.
Quote aggregator	These real time websites allow the consumer to enter their information once and receive multiple quotes instantly.
Insurance Mall	Similar to Quote Aggregator with extra function of online purchasing.
Direct Channel	Websites offer real-time quotes and instant online purchasing.
Virtual carrier	All activities are carried out on the web with little physical customer service centres.
Quote Mall	Gather information from a customer and then pass it to an appropriate agent to prepare a quote.
Agent Mall	Common websites shared by multiple agencies.
Consumer Auction	Consumers enter the information required for a quote and agents then bid for the business on the web.
Carrier Auction	Websites allow the agent to enter case information, and then the carriers bid for it.

Source: http://www.insurance.about.com/industry/insurance/library/weekly/aa04200a.htm

Finally, the size of the manufacturing enterprises also affects the business model. Large corporations with large user population are able to afford to purchase and/or license and therefore deploy expensive web applications at their own central enterprise portal. However, this is unlikely for the small and medium sized enterprises that are unable or unjustifiable to buy and/or license the web applications at their own web sites. Solutions must be sought to overcome these difficulties.

6.3.   FUNCTIONALITY OF WEB APPLICATIONS/CONTENTS

The functionality and purpose of web applications vary widely from one another. For example, some web applications allow manufacturers to sell directly to customers; to buy directly from suppliers; to negotiate directly with subcontractors; to support the customers directly; to find and work with business partners; to process business transactions; and to process technical transactions; etc. Therefore, the functionality has been used to classify their business models.

For example, Timmers (1998) classified Collaboration platforms as a business model. These web applications provide a set of tools and an information environment for collaboration between enterprises. These can focus on specific functions, such as collaborative design and engineering, or providing project support with a virtual team of consultants. Business opportunities are embedded in managing the platform (membership/usage fees), and selling the specialist tools (e.g. for design, workflow, document management).

However, the classification purely based on the functionality is hardly convincing. There are hundreds of different functions of web applications in practice and they cannot be classified as business models. In addition, the functionality such as collaboration of web applications does not specify how the revenue is generated. Therefore, it can be concluded that a classification scheme solely based on the functionality dimension is inadequate.

6.4.   VALUE CREATION

The Value Creation method is the main dimension of a business model. This can easily be observed from the investigations reported in the literature. Timmers (1998) proposed a systematic approach based on value chain de-construction and re-construction, to identifying architectures for business models. Value chain de-construction and re-construction is taken as identifying value chain elements, and identifying possible ways of integrating information along the chain. It also takes into account the possible creation of electronic markets. These can be fully open, that is, with an arbitrary number of buyers and sellers, or 'semi-open' that is with one buyer and multiple sellers (as in public procurement) or vice-versa.

Value chain de-construction means identifying the elements of the value chain. Nine typical value chain elements are inbound logistics, operations, outbound logistics, marketing & sales, service; technology development, procurement, human resource management, and corporate infrastructure. The first five are primary elements and the last four are support activities respectively.

Value chain re-construction is an integration of information processing across a number of steps of the value chain. The combinations are of the value chain elements involved. Possible architectures for business models are then constructed by combining interaction patterns with value chain integration. For example, an electronic shop is 'single actor'-to-'single actor' marketing & sales. A basic electronic mall consists of N times an e-shop. An electronic mall having a common brand offers many-to-1 marketing & sales (brand information is common across 'many' suppliers in the mall). An electronic auction where multiple buyers are bidding for the sales offer of one supplier brings together the sales of one supplier with the procurement of multiple buyers at a time, while combining the bid information from the multiple buyers.

6.5.   REVENUE GENERATION

The revenue generation dimension is naturally twinned with the value creation dimension. Schlachter (1995) identified five possible revenue streams for a web site. They included subscriptions, shopping mall operations, advertising, computer services and ancillary business. The emphasis was to show how would the revenue models existing in the brick and mortar scenario be exploited in a web based business.

Fedewa (1996) identified seven revenue generating business models. In addition to the revenue streams identified by Schlachter, Fedwa added timed usage, sponsorship, and public support as possible revenue streams.

Based on a qualitative analysis of the Internet based models pertaining to grocery and delivery of customer packages, Parkinson (1999) stressed the role of business affinities such as logistic providers in creating the value proposition.

The classification of business models based on the source of revenue is straightforward to understand in contrast with other classification schemes. One drawback is that a digital enterprise derives its revenue from several sources. In this case, it is very difficult and often confusing to use one revenue source to describe the business model of the digital enterprise. For this reason, other dimensions must be introduced.

6.6.   ACCESS CONTROL

The methods of access control have been used for understanding and classifying the business models of web applications. Registration and Subscription are two of the most widely used access control methods. Usually, there is free registration service, but sometimes, it is charged. Infomediaries provide content-based sites that are free to view and the user just needs to register first (other information may or may not be collected). Registration allows inter-session tracking of users' site usage patterns and thereby generates the data of greater potential value in targeted advertising campaigns. This is the basic form of infomediary model. The main source of revenue in this type of infomediary model is generated from advertisements. In addition, the infomediary itself obtains greater publicity.

Subscription is usually charged but sometimes for free. Users pay for access to the site. High value-added content is essential. Examples include Wall St. Journal, Consumer Reports. Generic news content, viable on the newsstand, has proven less successful as a subscription model on the web. An example is http://www.slate.com/. A 1999 survey by Jupiter Communications found that 46 percent of Internet users would not pay to view its contents on the web. Some businesses have combined free content (to drive volume and ad revenue) with premium content or services for subscribers only.

In the literature, the Registration and Subscription methods of access control are classified as two different business models (http://ecommerce.ncsu.edu/topics/models/models.html). Both registration and subscription can be a source of revenue. But they do not seem to define the value creation sufficiently to be qualified as business models. In this sense, registration and subscription are classified as two methods for controlling the user access, which in turn constitutes a key dimension of the business model of digital enterprise.

In addition, there is one issue closely related to access control that is the access security and confidentiality. Although largely ignored in the literature, security is considered as a unique and important component of business models of digital enterprises. This issue, if not handled appropriately, could be a dominating factor determining the success or failure of a digital enterprise. For example, the security and confidentiality must be absolutely guaranteed in a third-party virtual marketplace.

6.7.   TOWARDS A COMPREHENSIVE FRAMEWORK

There are relatively comprehensive frameworks for investigating business models of traditional enterprises. An example is discussed at the web site (http://www.howarddowding.com/). However, systematic efforts have just started in search of such frameworks.

Afuah and Tucci (2000) conceptualized a business model as a system that is made up of components, linkages amongst the components, and dynamics at different levels. How well a system works is not only about the function of a particular type of components, it is also about the function of the relationships amongst the components. Thus, if the value that a firm offers to its customers is low cost, then the activities that it performs should reflect that.

In addition, the relationship between the business model and its environment is also critical for determining the quality of a business model. A good business model alwaystakes advantage of any opportunities in its environment while trying to dampen the threats from it. However, the components and linkages of a business model do not last forever. Managers often have to change some components or relationships before competitors did that. In some industries, firms have to keep renewing their business models. They have to cannibalize themselves before someone else does so. Afuah and Tucci defined *dynamics* as those actions associated with changes, initiated by a firm to pre-empt competitors or to fend them off, or in response to any other opportunities and threats.

Barua and Whinston (1999) proposed a four-layer framework for measuring the size of the Internet economy as a whole. The Internet *infrastructure layer* addresses the issue of the backbone infrastructure required for conducting business via the net. Expectedly, it is largely made up of telecommunication companies and other hardware manufacturers such as computer and networking equipment. The Internet *applications layer* provides support systems for the Internet economy through a variety of software applications that enable organizations to commercially exploit the backbone infrastructure. Over the years, several applications have been developed, addressing a range of issues from web page design to providing security and trust in conducting various business transactions over the net. The Internet *intermediary layer* includes a host of companies that participate in the market making process in several ways. Finally, the Internet *commerce layer* covers companies that conduct business in an over all ambience provided by the other three layers. The Internet *infrastructure layer* and the *applications layer* play a crucial role in moderating and trend setting the growth of Internet economy.

Mahadevan (2000) argued that a business model is a unique blend of three streams that are critical to the business, namely the value stream for the business partners and the buyers to identify the value proposition for the buyers, sellers and the market makers and portals in an Internet context. The revenue stream is a plan for assuring revenue generation for the business. The logistical stream addresses various issues related to the design of the supply chain for the business. The long-term viability of a business largely stems from the robustness of the value stream. Furthermore, the value stream in turn influences the revenue stream and the choices with respect to the logistical stream.

6.8.   SUMMARY

This chapter has briefly reviewed the recent efforts on investigating business models for Internet/web-based enterprises. Several dimensions where business models can be constructed have been highlighted. They must be applied at the same time, instead of being used in separation. On the one hand, a wide variety of business models have emerged. Some of these models are essentially an electronic re-implementation of traditional forms of doing business, such as e-shops. Others become feasible only because of the openness and connectivity of the Internet. On the other hand, it is still a mystery of what constitutes the business model of a Category II digital enterprise. It is still unclear how the existing traditional business model is related to and consequently affected by the business models of Category I digital enterprises when they are being combined.

Several comprehensive frameworks for understanding or evaluating business models of digital enterprises have been briefly discussed. While the efforts continue, following observations can be obtained:

- A business model should define the way that values are created for all business actors involved in the business including the Application Service Providers, the Application Users, the Application Subscribers/Licensees, and the Data Providers, etc.
- A business model should define the way that revenues/benefits are created for all business actors involved in the business.
- One unique feature of business models for digital enterprises is about how security/confidentiality is guaranteed between all business actors. This dimension has not been incorporated into the above theoretical frameworks although being considered in practice.
- Access controls should be defined in the business model for any digital enterprises.
- A comprehensive business model should define an architecture for the flow of products, services and information, including a description of various business actors and their roles;
- A business model itself does not yet provide information about how it will contribute to realize the business mission of any of the companies who is an actor within the model. Therefore, it is useful to incorporate the marketing strategy of the business actor under consideration into its business model to form what is usually referred as "marketing models". Through this combined model, it becomes possible to assess the commercial viability and answer questions like: how is competitive advantage being built, what is the positioning, what is the marketing mix, and which product-market strategy is followed, etc.
- Information and communication technology enables a wide range of business models. However, the capability of the state-of-the-art technology is just one criterion in model selection and technology. It provides no guidelines for

selecting a model in commercial terms at all. Alternatively, guidance to technology development can come from the definition of new models.

- Although the above frameworks do not exclude the business models for Category II of digital enterprises, majority of them focus on those of Category I digital enterprises. Therefore, they do not adequately address the interface or integration between the business models of the digital business with those of the existing traditional one.

- Finally, it is inadequate to use just one or two dimensions such as revenue creation and access control to define business models of digital enterprises. A systematic approach or framework would integrate several key dimensions. Parameterization along these dimensions would lead to more innovative business models.

7

SEARCH ENGINE OF WEB APPLICATIONS IN PRODUCT INTRODUCTION

There have been widespread interests in the development and application of web-based systems to support decision-making activities in product introduction/innovation process. An increasing number of web applications are emerging and a large number of practitioners are keen on trying these remote decision support systems (TeleDSS) through the web browsers. In the meanwhile, it becomes increasingly difficult to surf for appropriate web applications on the Internet with general-purpose search engines.

This chapter discusses the design, development and operation of a special-purpose search engine – WAPIP. It provides a web site for software vendors and researchers to register their web applications with the "wapip" search engine. It also provides facilities to support practitioners in product development to search rapidly for the right web applications that are suitable for solving their problems. All web applications entered into this WAPIP web site must be solicited before they are made available to practitioners by qualified solicitors or experienced users. It is expected that WAPIP can improve efficiency and effectiveness significantly for supporting product development/introduction in the extended enterprises.

The rest of the chapter is being organised as follows. Section 7.1 will outline the need for a special search engine like WAPIP. The design and development issues of WAPIP will be discussed in Section 7.2. Section 7.3 will focus on the implementation details. Section 7.4 will describe the working procedure and mechanisms. Directions for further developments will be presented in the last section to conclude the chapter.

7.1. THE NEED FOR SPECIAL-PURPOSE SEARCH ENGINE

The need for a special-purpose search engine in the field of product development/introduction can be seen from the following aspects:

- The number of web applications in product development has increased over the recent years, and is expected to increase dramatically in the near future.
- General-purpose search engines are inadequate to help practitioners to search for web applications most appropriate for their tasks.

7.1.1. Dramatic Emergence of Web Applications

One of the first and most significant initiatives in the development and application of web-based systems in product design and manufacture is the substantial American research project the MADE (Manufacturing Automation and Design Engineering) program. MADE was a DARPA (Defence Advanced Research Projects Agency) program first initiated in 1992 and eventually completed in 1996. The MADE program supports research, development and demonstration of enabling technologies, tools, and infrastructure for the next generation of design environments for complex electro-mechanical systems. This program involved a number of major research centres/groups, resulting in valuable publications at conferences, journals, and on the Internet (Cutskosky et al, 1996; Petrie, 1996; Whitney et al, 1995; Bryant et al, 1995; Will, 1996). This program concerns with the comprehensive information modelling and the design tools needed to support rapid design of electro-mechanical systems. This program emphasises the notion of "tag team" design in which each designer performs the functions that he or she is most expert at while leaving behind in a design information web enough information for other designers to pick up wherever the others left off. MADEFAST was a demonstration of this approach conducted by several research groups who collaborated in the design and manufacture of a prototype sensor array aiming system. The MADE program continues as RaDEO (Rapid Design Exploration and Optimisation) program (RaDEO, 1997). Since then, there have been rapid developments in web applications in collaborative product development.

Chapter 2 has provided more detailed reviews on the web applications of product development in the following three categories: (1) Individual web applications that are more conducive to product design and manufacture; (2) Web applications that are especially designed and developed to facilitate and support group or team work in collaborative product development; (3) Inter-operation between individual decision support systems. Implementation issues are discussed separately in Chapters 4 and 5.

7.1.2. General-Purpose Search Engines are Inadequate

There are many general-purpose search engines. Yahoo!, WebCrawler, AltaVista, Hotbot, InfoSeek, Lycos, Excite, Northern Light are some examples commonly used all over the world. While they are generally very useful in searching for information on the Internet, they are inadequate in supporting product development practitioners (Spink et al, 1999). There are several reasons for that. Firstly, as a core element of search engines, repositories for general-purpose search tools always cover a wide variety of domains or subjects. The specialized information, especially in product development, always takes a small partition in those repositories. It is very difficult to extract such a narrow segment from a mountain of information.

Secondly, since general-purpose search engines supply services for all kinds of users. For specialist users like those involved in product development, it is hard to submit a suitable search item that exactly describes their original intention. Therefore, they often encounter a great number of results, of which, the majority are possibly irrelevant to their tasks.

Thirdly, many search engines build up their repositories by matching parts of documents with keywords. The results are often not very relevant and the results of a search query tend to be too large for a user to filter within a short period of time.

Finally, for individual web application vendors, they are reluctant to register themselves with those general-purpose repositories. This is because important information can seldom survive through a general form as required by search engines. As a result, the users hardly seek them just by using the general search criteria.

The limitations described above could be explained through an example. Suppose that a project team member is interested in using web-based Quality Function Deployment (QFD) system to support the product design specification. Two popular search engines, namely "Yahoo!" and "AltaVista" are used in the experiment. The keyword "QFD" is used. It was hoped that appropriate web-based QFD systems can be found on the Internet through this search query. However, the results are more a frustration than a satisfaction. Several observations can be made from the experiment:

- Numerous results are generated. Table 7.1 shows the query results from the two general-purpose search engines, namely, Yahoo! and AltaVista respectively. In the Yahoo! result statistics, there are no categories about "QFD". However three corresponding web sites are caught. Over three thousand web pages are located, while neither related news nor net events come out. The number of items resulted from the keyword search with AltaVista is enormous, as can be seen from Table 7.1(b).

- Time-consuming to filter relevant information. The users must browse through thousands of query results, item by item, to recognize whether results are practical or not. This process is extremely tedious.
- Few resulting hyperlinks point to web applications. Unfortunately, only a few of the hyperlinks resulted from the experiment provided direct link to QFD systems on the Internet. Most hyperlinks are web pages promoting or explaining the QFD technology. Some links promote standalone QFD computer software systems.

Table 7.1 Search results from different search engines for keyword "QFD"

(a) Yahoo! results		(b) AltaVista results	
Categories	0	Web Pages	8234
Web Sites	3	Software	258270
Web Pages	3252	Capture	112
Related News	0	What is …	572975
Net Event	0	Designer	847795

(The experiment was conducted in November 1999. A slightly different result may be obtained today. But we expect the difference to be minimal.)

It could be drawn from the above experiment that general-purpose search tools are not adequate to support product development practitioners to find suitable web applications. This leads the authors to the development of a special-purpose search engine where software vendors can register their web applications. Practitioners involved in product development can easily retrieve the web applications most suitable for their development tasks.

7.2.   SYSTEM DEVELOPMENT: OBJECTIVES AND APPROACHES

7.2.1.   Working Scenario

Before analysing system architectures in detail, it is necessary here to outline the working scenario that this research work aims to address. Assume that a team is established for a new product development project. The scope of the project varies widely, depending on the nature of the project. The project can be as focused as a Design for Assembly (DFA) analysis. In this case, one web application for DFA analysis would be sufficient. However, the project can be very large, involving a number of work packages. Each work package requires one web application to facilitate the decision-making process. Again, let us assume that the project team or member needs to search for a web application on the Internet, which are suitable for the project or the work package. Given the tight project schedule, team members would not entertain the surfing with general-purpose search engines as described in the preceding section. What this project aims to achieve is to establish a web site where they can find the web application(s) they want to use for the project.

7.2.2.   Objectives of the Special-Purpose Search Engine

A special-purpose search engine aims to provide a matching mechanism on the Internet between the vendors and the users of web applications in the context of product development and introduction process. There are three specific objectives to be achieved:

- To develop a vendor and its web applications registration module which creates an access to the system service for collaborations;
- To develop a system solicitation module which is in charge of filtering the qualified web applications into system repositories;
- To develop a search module which enables engineering project teams/customers to seek for the qualified collaborative partners through an express channel.

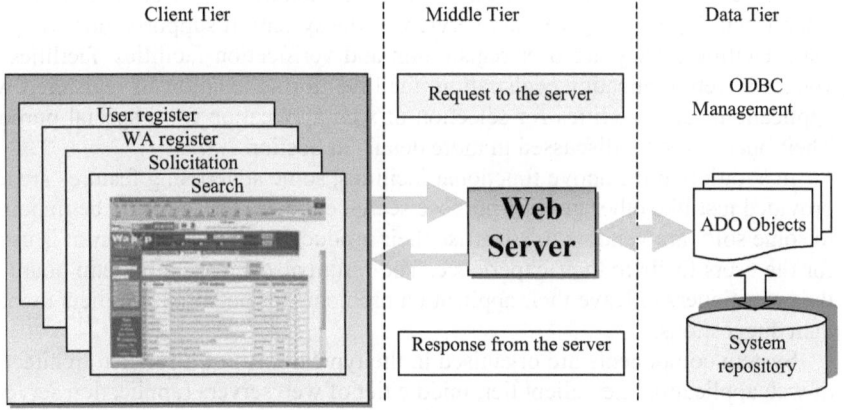

Figure 7.1 Overall of the special-purpose search engine – WAPIP.

7.2.3.   Approaches to Building Search Engines

There are two general approaches to building information searching facilities on the Internet through web browsers (Gould, 1998). The first approach is by subject directories. It gathers many different home pages of web sites but does not go deep within each site, and browses or searches a small database of titles and annotations of web sites. The second approach is through search engine approaches. By this method, fewer web sites are visited but hyperlinks are followed many levels deep within each site. An enormous database of web sites that includes full text of each resource is established.

In comparison, the two approaches differ in the way of collecting and storing information in their database. The former emphasises on the breadth of information, while the latter sacrifices the breadth for depth. Usually, it is difficult to determine the extent that they are being used by a search tool respectively. The trade-off between breadth and depth must be achieved by all search tools in order to perform effectively, given the complexity of the medium, the limited storage capacity, and retrieval speed of even the most advanced equipment.

The approach adopted here for developing this special-purpose search engine basically follows the first approach of subject directories. However, further developments have been planned to incorporate search engine techniques after the first prototype is demonstrated.

7.3.   OVERVIEW OF *WAPIP* SYSTEM

7.3.1.   System Components and Architecture

After the initial round of research and development, a web site has been established. Figure 7.1 gives an overview of the system. It supports four groups of main facilities. They are user registration and verification facilities, facilities for vendor's web application registration, facilities for solicitation of registered web applications, and facilities for selection of web application for a special purpose. Their operations are discussed in more details in Section 7.5.

In addition to the above functional facilities, some advertising features are also provided just like other general-purpose search engines. These would be important to some software vendors to advertise their products. A discussion forum is useful for the users to share their experience. The forum also acts as a bulletin board for the practitioners to leave their application problems for potential suppliers to make their suggestions.

System components are organised in the typical form of three-tier architecture of web application, i.e., client tier, middle tier of web servers (application servers), and data tier. Application clients include client browsers with which the customers connect to the system web Server. The Web Server (together with the Application Server) is at the center of the system. It provides the objects used by the client and controls the information communication between the other two tiers. The objects will provide three main functions in the application, namely, registration, search and solicitation. Data source and ODBC management provides the repository of web applications, etc for the system. ODBC management acts as the information broker between the data source and the application, which retrieves data from database by the Open Database Connectivity (ODBC) and transfers data to client by the Active Data Objects (ADO).

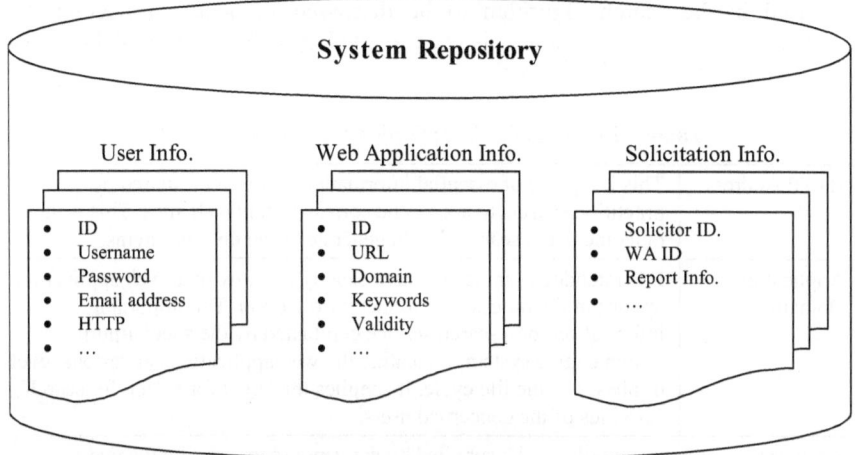

Figure 7.2 System data model.

7.3.2. System Data model

According to aforementioned discussions, the repository is the heart of an integrated special-purpose search engine management strategy. It provides a space, in which information about web applications is entered, kept up to date, and made available to everyone who needs it (McClure, 1992). Therefore, the system data model, as the core element of the repository, should be the mechanism for defining, storing, accessing, and managing all the information about a web applications, its sharing data, supplier contact items, and so on. Basically, both the vendors and solicitors are the users who usually supply the information into the repository. The practitioners are the users who consume or use the information.

The following key information, also indicated in Figure 7.2, is included in the WAPIP database:

- User information. Details of four types of users must be maintained in the system. For example, their contact details must be made available to the user group so that they can communicate and discuss with one another. The email address is essential. Their usernames, passwords, validity, etc. are also required during the login process.
- Solicitation reports. At present, solicitation is rudimentary. In fact, this information is recorded in the web application registration table as a field of validity. In the future, it is necessary to provide comprehensive solicitation reports, so that practitioners can make their better choices from the WAPIP repository.
- Web application registration information. The main content of the WAPIP repository is the information about web applications. Each web application must have its name, plus the details shown in Table 7.2. All these items are

used in the search algorithm to be discussed in detail in Section 7.4.4. However, these items should also be provided in such a way to facilitate visual browsing by practitioner users.

Table 7.2 Details of web application registration information

HTTP Address	This is a piece of essential information that will be shared by practitioner users to access the corresponding web sites. This is the outcome that a search should generate based on other items.
Application Domains	It is intended to indicate where exactly an individual web application applies in the process of product introduction. The important information for a search service conducted by the special-purpose search engine system, indicating the web applications at various levels or phases of the life cycle. It supplies the key information for search activities of the concerned users.
Keywords	Keywords can be supplied by the vendors to describe the main functions of the web applications, so that they can be quickly retrieved from the repository or easily undertood without reading lengthy descriptions.
Main Description	These are relatively longer descriptions about web applications. Although these sentences can be used in the search algorithm, their main purpose is to help the practitioner and solicitor users to better understand the capability and limitations of the web applications.
Validity	This information is the outcome of the solicitation process. If a web application is successfully evaluated by a solicitor, it is assigned a positive validity. Otherwise, it has a negative validity. Invalid web applications are not made available to the practitioner users.

7.3.3. System Implementation and Deployment

There are several techniques that can be used for implementing the WAPIP system. For example, server-side program technology such as CGI Common Gateway Interface) is a traditional way of submitting and retrieving information to and from the web server. In addition, the combination of client-side Java Applet, Java Script, VB Script and ActiveX Objects with server-side components such as Java servlets and ActiveX components can achieve the same objectives of client-server interactions. These techniques usually involve intensive programming (e.g. CGI codes and client components). Resulting clients can sometimes be very "fat" that requires a long time for downloading.

Active Server Pages (ASP), which is considered as an enhancement of common CGI applications, is a server-side scripting environment that designers can use to create dynamic web pages or build powerful web applications. It improves the capabilities, facilities and compatibility of common CGI applications. What is more, ASP pages can call ActiveX components and Java applets to perform tasks, such as connecting to a database or performing a business calculation (Http://www.learnasp.com/). With ASP, designers can add interactive content to web pages or build entire web applications that use HTML pages as the interface to customers. For these reasons, ASP is used for implementing the WAPIP prototype. Experience shows that this is an efficient technique for achieving the desired functionality. The downloading time is much shorter than the "fat" client techniques.

7.4.   SYSTEM OPERATION

There are three user groups in total. The first group is a group of software developers or vendors who would like to register their web applications with the search engine. The second group includes those practitioners who would like to choose and thereafter use particular web applications to solve their problems in the projects. The final user group is the search engine provider who would administer all the facilities and also verify the registration information before a web application can be made available to the practitioners' community. Accordingly, three main components (web pages) have been provided to support these three groups . They are web application registration facilities, web application search facilities, and web application solicitation facilities respectively. Of course, additional facilities are available for managing the users.

7.4.1.   User Registration and Login Controls

The intended user community of the WAPIP web site is expected to be small compared with those of general-purpose search engines that do not usually require the users to register. The sample screen of the user registration is shown in Figure 7.3. In contrast, users must register themselves with WAPIP by providing necessary information. Usernames, passwords and user's email are very important because they are used in the future for login purpose and for the administrator to contact the users through email. For example, if a user forgets about the username and password, the email address can be used to get the information.

The user registration must be verified before the user can use the WAPIP system. User verification is accomplished by the system administrators. An email acknowledgement is sent to the new users upon successful verification. Should a user forget about the username and password, a facility is available to arrange the registration information to be sent to the email address as appeared in the original registration. It is our intention that only research web applications are made available to unregistered practitioners. Commercial web applications are accessible by registered users only. Registered users would enjoy other facilities, such as virtual teaming, to be incorporated into the system in the near future. One user may be registered as all three types of users (vendor, solicitor, and practitioner).

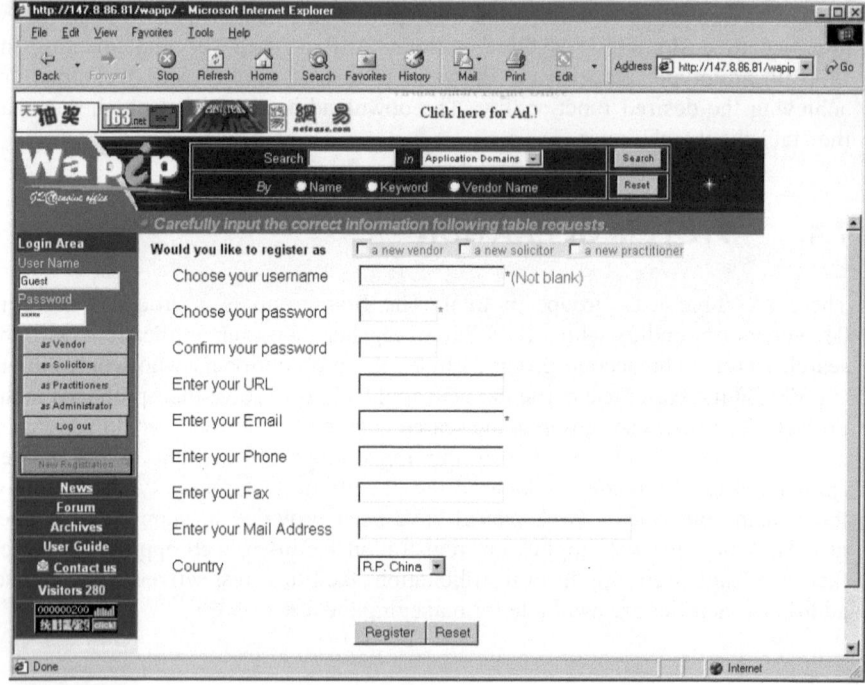

Figure 7.3 Login control and user registration.

7.4.2.   Registration of Web Applications

If a user is registered and logged in as a vendor, he or she can register their web applications in the WAPIP system. Existing web applications can be modified. Even vendor' details can be changed at this stage. Figure 7.4 shows some facilities for registering web applications.

It is essential that the correct URL (Universal Resource Locator) of the web application must be provided upon registration. This hyperlink will be used by the solicitor to evaluate the web application. It is recommended that this URL should point directly to the web pages of the web application.

As optional, application domains, keywords, general descriptions, and the names of the web applications should be properly edited and provided, they have initial effects on the practitioner users to decide if they are interested in trying the hyperlinks.

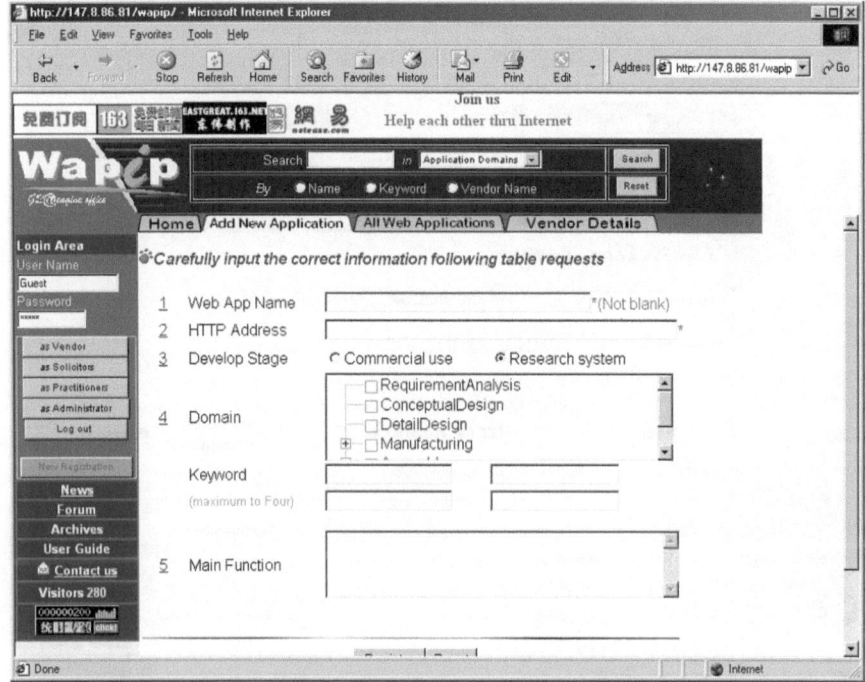

Figure 7.4 Registration of web applications.

7.4.3. Solicitation of Web Applications

One of the main objectives in developing this special search engine is that all the web applications registered must be highly relevant to applications in product design and manufacture. Thus, the searching time and efforts by the practitioners can be minimised. In order to achieve this objective, it is necessary to solicit the registration before a newly entered web application is made available to the practitioners' community. Figure 7.5 shows some of the solicitation facilities.

One of the main objectives of solicitation is to ensure the high relevance of the web application against its application domain, and the hyperlink provided points directly to the web application rather than promotional materials only. Only those web applications that are successfully solicited will be made available to the practitioners' community. Those fail the solicitation will not be published in the WAPIP database.

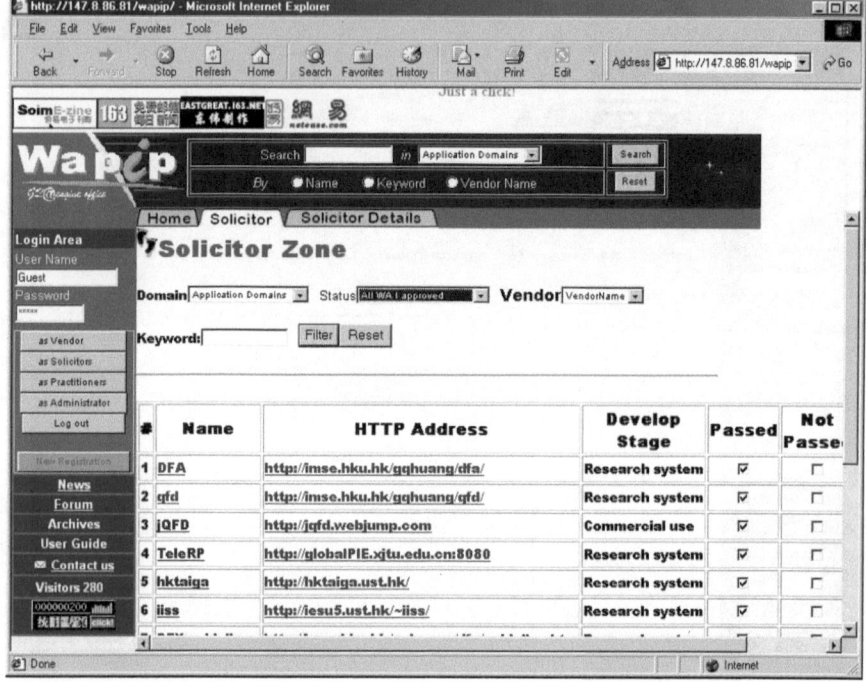

Figure 7.5 Solicitation of web applications.

There are two basic questions here. One is who should be responsible for soliciting web applications. The other question is how should the solicitation be carried out. It must be noted that the system administrator should not be tasked with solicitation. Only those person who have the specialist knowledge about the application domain, or those who have extensive experience in the application domain, are qualified as solicitors. WAPIP provides a registration facility for those who want to act as solicitors. However, there are no qualification elements at present.

At present, the solicitation is rudimentary by simply checking "pass" or "fail" without giving a full report. Metrics must be identified for solicitation. A Checklist or guidelines need to be developed for the solicitation process. Resulting reports and comments should be made available to both the vendor and practitioners.

Figure 7.6 Searching for appropriate web applications.

7.4.4.   Searching for Web Applications

Figure 7.6 shows an overview of the search facilities while Figure 7.7 shows the search algorithm. There are several modes for choosing appropriate web applications. Firstly, the practitioners can specify an application domain. All the web applications registered under this domain will be retrieved to the user. This is a very effective search mechanism because most practitioners know the applications domains of their projects or work packages. However, an issue arises, that is what are application domains and how should they be best represented in WAPIP. An ideal scheme is a roadmap of decision-making activities in the product introduction process. Reinders (1995) uses a tree structure to group design assistants. However, different researchers and practitioners may dispute the existence of such a generic roadmap.

The second search mode is by vendor name if the user knows. All the web applications of the named vendor will be retrieved for the user to consider. This search mode is useful if the user is familiar with software vendors.

The third search mode is by the name of the web application. This only applies when the user knows or has used the web application before.

The fourth search mode takes place within the keyword fields of web applications. Up to four keywords are allowed upon registration to describe the web applications very concisely. Typically, keywords are about the functionality of the web applications. This allows another quick "find".

In case that no search mode is defined, a general search procedure takes place. If no keyword is defined in the text box, then all web applications are listed, that is, no search takes place. If a keyword is specified, then a hierarchical search algorithm, shown in Figure 7.7, is followed to find the matching web applications. The steps of search are as follows:

- First, a keyword is used to match the keywords of web applications in the database. If a matching is found, the corresponding web applications will be short-listed in the result.
- Second, the keyword is then used to match the names of web applications provided at the time of registration. If a match is found, the corresponding web applications will be short-listed in the result.
- Third, the keyword is used to match the descriptions of application domains. If a matching is found, the corresponding web applications will be short-listed in the result.
- Finally, the keyword is used to match the general descriptions of web applications. If a matching is found, the corresponding web applications will be short-listed in the result.

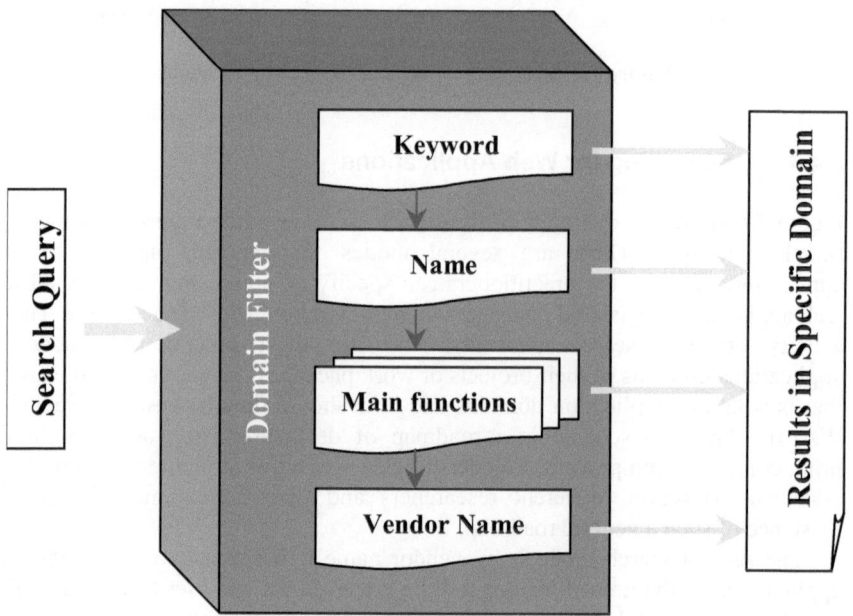

Figure 7.7 Hierarchical search structure and implementation facilities.

The search engine returns none or several web applications. In case of multiple results, the user must make the final choice using other criteria or by browsing through the list. At present, the search engine does not include any algorithms for calculating the matching indices to rank the order of the resulting web applications. This would be desirable when more software vendors register their web applications with the search engine repository. More sophisticated searching algorithms such as Vector Space Modelling (Berry and Browne, 1999) would be incorporated in the further development.

The practitioner user can build up his/her favourite web applications based on search results. This mechanism allows the user to bookmark the web applications (s)he often uses or most likely to be interested in. Once added to the favourite list, the web applications can be called directly without search.

7.5. SUMMARY

A new special-purpose web site, WAPIP, has been presented in this chapter. It can be used by the following three groups of professionals:

- For a software vendor or a research group involved in developing new web applications relevant to product design and manufacture, WAPIP can be used to register the web application.
- An experienced user of web applications in product development can act as a solicitor to evaluate web applications registered with WAPIP for their suitability for supporting product design and manufacture before they become available to the practitioners.
- A practitioner involved in new product development can use WAPIP to search for a web application most suitable for supporting the design tasks without the hassle of wasting time and efforts like using general-purpose search engines.

The development and implementation of the search engine is technically straightforward. However, the real challenge now lies in the collection of sufficient number and variety of web applications in product design and manufacture into the repository, so that meaningful experiments can then be conducted in industrial settings. To achieve this objective, we need participation from all the researchers, product development practitioners, and software developers/vendors.

More sophisticated searching algorithms should be incorporated into the search engine, so that searching results are rank ordered in terms of their relevance. The Vector Space Model (Berry and Browne, 1999) is being investigated as a candidate methodology.

In addition, a more comprehensive model of the product introduction process is required to indicate clearly where and what decisions a particular web application is applicable. The resulting model is expected to provide a better roadmap for applications, as used in the WAPIP system. This would benefit both the software developers and users.

As already mentioned previously, the solicitation needs to be further improved and strengthened. There are basically two options. One is to loose the solicitation, allowing software vendors to register their products in the WAPIP system with little or without solicitation. This option is vulnerable to web vandalism where irrelevant information is entered into the WAPIP system. The other option is to tighten the solicitation, just like referees reviewing the manuscripts submitted for publication in journals. This is preferable if there are enough voluntary and knowledgeable solicitors.

Finally, the ultimate objective is to extend this simple search engine into a virtual office for product development. Therefore, a variety of virtual teaming facilities must be incorporated. Although the methodology has been established from our previous research, its practical implementation and operation depends on the successful registration of sufficient web applications.

8

WEB-BASED DESIGN FOR MANUFACTURE AND ASSEMBLY

Design for Manufacture and Assembly (DFMA) has received widespread attention from both the research and practitioner communities. As a result, there have appeared a number of well-known techniques, accompanied by computer-aided systems. This chapter is concerned with providing DFMA techniques on the Internet. The DFMA analyst user simply uses an HTML (HyperText Markup Language) browser to connect to an appropriate DFMA web site and starts analysis immediately. If necessary, a subscription is done in advance or on-line. With this web-based DFMA, no installation is needed at the client (analyst) side, nor is maintenance which is done at the server side.

The aim of this chapter is to investigate the issues related to the development and application of Design for Manufacture and Assembly on the Internet. First, an experiment is reported on the implementation of a WWW (World Wide Web) homepage for a well-known Design for Assembly technique. An experiment is conducted to show how a well-known Design for Assembly (DFA) technique can be converted into a web-based version which is functionally equivalent to its version on a standalone workstation. DFX involves both product and process design decisions and therefore is a typical scenario or episode of collaborative product development. Important insights will be sought from the experiment. Once sufficient insights are gained from this web-based DFA research, findings may be extended to other scenarios of collaborative product development on the Internet.

Implications from this experiment are then discussed under the four categories: client-server architecture for collaborative Design for X (DFX), web-based framework for developing DFX tools, web-based framework for integrated utilisation of DFX tools, and web-based framework for collaborative product realisation. The web-based client and server architecture is found to be attractive for collaborative DFMA. The client-side web scripting can be exploited to develop generic frameworks for developing and applying different Design for X (DFX) techniques, more importantly, in an integrated way. In addition, web-based DFX techniques provide more opportunities for integration with other decision-support systems such as Computer Aided Design (CAD), Computer Aided Process Planning (CAPP) and Computer Aided Production Management (CAPM) in the product realisation process. However, issues such as interactivity and security remain to be addressed.

8.1.   WEB-BASED DFA ARCHITECTURE

As a feasibility study, an experiment is conducted to implement a well-known DFA tool on the Internet. The objectives of the experiment are to investigate into (1) various approaches to developing web-based DFA tools; (2) whether the current web-based technology is sufficient for developing web-based DFA tools that are functionally equivalent to those implemented on standalone workstations; (3) the advantages and disadvantages of the web-based DFX environment; and (4) implications for further development.

There are two major approaches to developing comprehensive DFMA techniques. One is to compile DFMA guidelines. It is relatively straightforward to implement DFMA guidelines on the web. Figure 8.1 is an example. DFMA guidelines are first indexed according to primary manufacturing processes as implemented by the drop-down Combo box. Product features and secondary process characteristics are used to identify the most appropriate guideline(s). For example, if the "casting" is the primary process, then the web page corresponding to this process is rendered at the browser. This page is static in the sense that its contents and formats do not change with the user actions. It is the user who browses through the page to locate appropriate guidelines and to follow them during the design or redesign activities.

Though useful, there are serious limitations with the guideline approach. Boothroyd (1996) has been a critic for a long time against the inefficiency of this approach. The recognition of the limitations was the beginning of the development of systematic DFA techniques. As a result, the Boothroyd-Dewhurst DFA (Boothroyd et al, 1994) and Lucas DFA (Swift, 1981) are among the best known of such techniques. Both have actually been supported by computer software packages. Static web pages are no longer sufficient to provide the functionality equivalent to these computerised DFA systems. A simple reason is that the pages must be interactive to user operations and the contents of a page must change with the user actions. Fortunately, the WWW technology has moved on to a stage where active pages become reality.

The Lucas Design for Assembly technique has been chosen for the experiment. The experiment has been carried out based on the Microsoft Internet Explorer. Visual Basic has been used for client-side web scripting. Web-based DFA system can be separated into two parts: server side and client side, as shown in Figure 8.2. The HTML files and initial database are stored in server side. The Internet Explorer browser, which is the software that runs on the client side, requests the web pages and other information from the server. The server sends them through Internet. Once the files arrive at the client site, the browser reconstructs and displays them on the screen as own files.

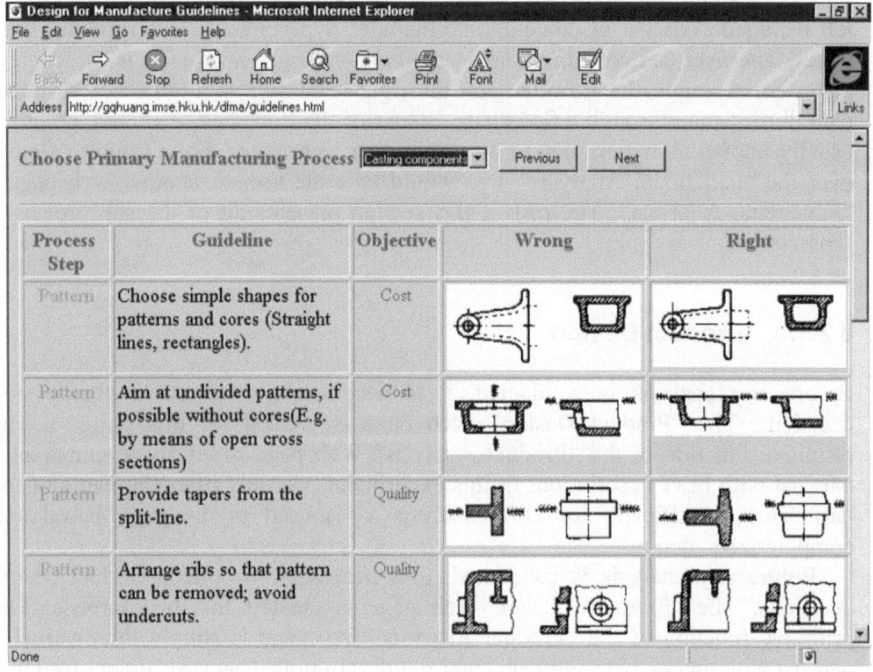

Figure 8.1 Static web page for DFMA guidelines.

A number of standard ActiveX controls are used. They are embedded in the web pages. Each of them corresponds to one analysis process during the DFA procedure. Some standard ActiveX controls are used to build the controls, such as Data Control, SSTab, MsFlexDBGrid and so on. All of these controls have been made into a cab file. When the user visits the site, he will be asked to download these necessary ActiveX controls from the cab file, as well as the database. The Web-based DFA system will check whether the client has these ActiveX controls and automatically decide whether and how to download them to client side. After downloading, these controls licensed in client side, dealing with the analysis procedure. With these controls, higher system interactivity can be achieved.

The DFA analysis database is downloaded from the server and stored in the client side. Anybody without permission has no access to the analysis data to reach a higher security level. Furthermore, the data is processed at the client side, which can decrease the amount of information exchanging between client and server, then increase the interactivity.

8.2.   WEB-BASED DESIGN FOR ASSEMBLY

Figure 8.3 shows the homepage of the experimental web-based DFA tool. On the left hand side is a list of contents that includes hyperlinks to corresponding web pages. The user can read them as a book. On the right hand side is the content page. For example the flowchart on the right-hand side of Figure 8.3 gives an overall road-map through a flowchart. Blocks of the flowchart are active pointers, usually known as hyperlinks, to corresponding web pages. For example, a click over the "Functional Analysis" area would take the user to another web page - "Functional Analysis". The rest of this section presents all of the sub processes separately.

8.2.1.   Product Design

Before any analysis is conducted, it is necessary to collect product design decisions. The "Product Design" Web page is created for this purpose. The primitive function of the "Product Analysis" Web page is simply to construct a part list with brief description. In this experiment, we only offer the name of the part. This is sufficient for DFA analysis as is used in the paper-based and computerized versions.

Before any analysis is conducted, it is necessary to collect product design decisions. The "Product Design" web page is created for this purpose. The primitive function of the "Product Analysis" web page is simply to construct a part list with unique part numbers and brief description. This is sufficient for DFA analysis as is used in the paper-based and computerised versions.

8.2.2.   Web-Based Functional Analysis

Once product data are collected, Functional Analysis is the first step in Lucas DFA analysis procedure. A very effective flowchart is provided to assist the users for Functional Analysis. This flowchart questionnaire can be easily implemented in the "Functional Analysis" Web page as shown in Figure 8.4.

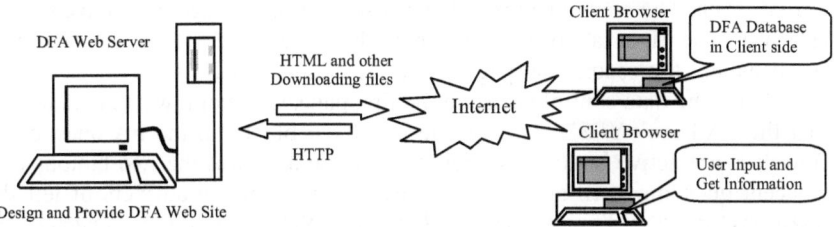

Figure 8.2 Web-based DFA architecture

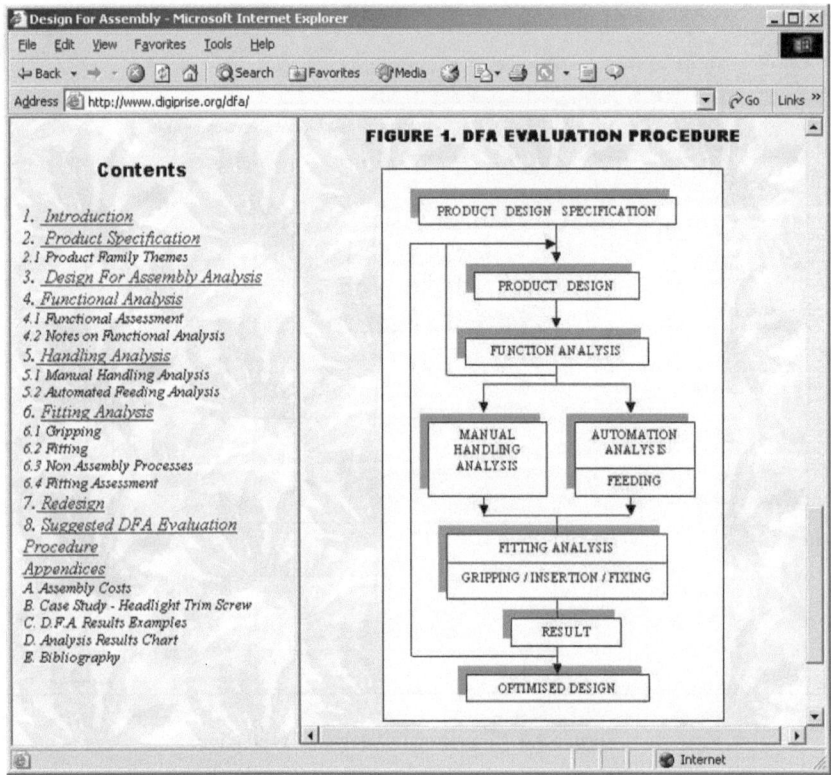

Figure 8.3 DFA procedure web page

Functional analysis is usually carried out to one part at a time. One data control is used to switch among different parts until all the parts are considered. At the end of Functional Analysis, parts are divided into "A - essential" and "B – non-essential" parts. A design efficiency index is then calculated based on the number of "A" part and the total number of the parts in the subject product. The main element of this "Functional Analysis" page is an ActiveX component coded in

Visual Basic. The component mainly consists of a multi-page tab ActiveX control (SSTab). The first tab page is for recording the results from the Functional Analysis and the second tab page is the Functional Analysis (FA) questionnaire flowchart. The data control is also in the first page. One point worth mentioning is that the <All are "YES"> and <Some is "NO"> blocks in the FA questionnaire flowchart are active buttons. A click on a button activates the two buttons of the adjacent question. <All are "YES"> means that all of the answers of left three questions are "YES", otherwise, <Some is "NO"> should be selected. This process continues until a final decision on whether a part is essential or non-essential is reached and recorded.

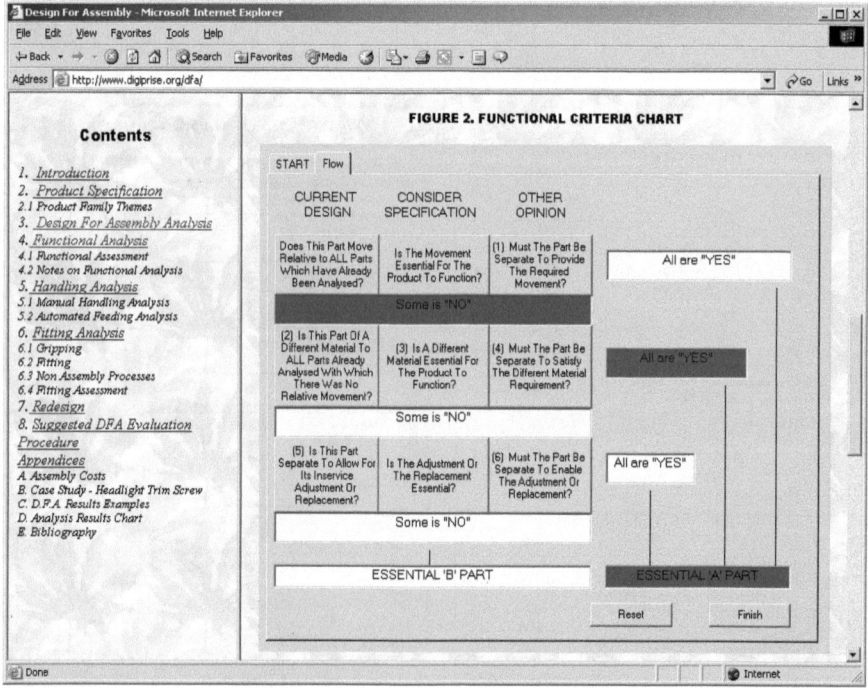

Figure 8.4 "Functional Analysis" Web page.

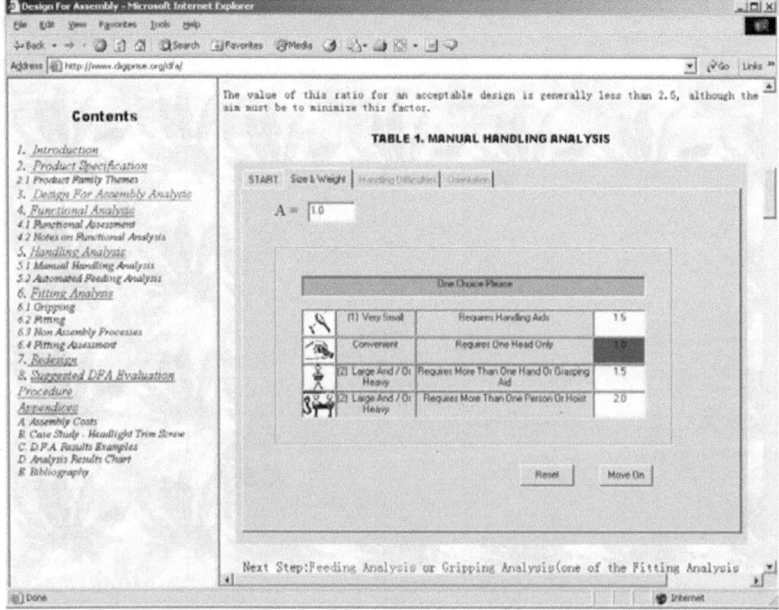

Figure 8.5 "Manual Handling Analysis" Web page.

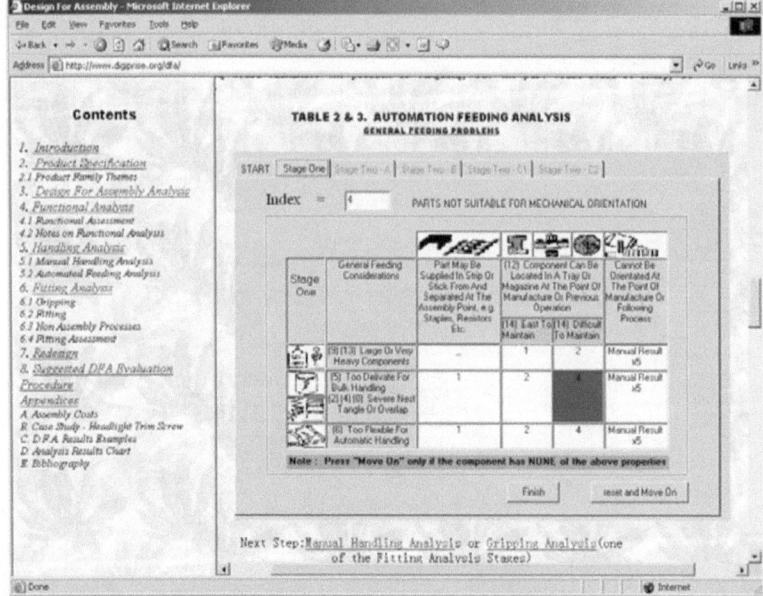

Figure 8.6 "Automated Feeding Analysis" Web page.

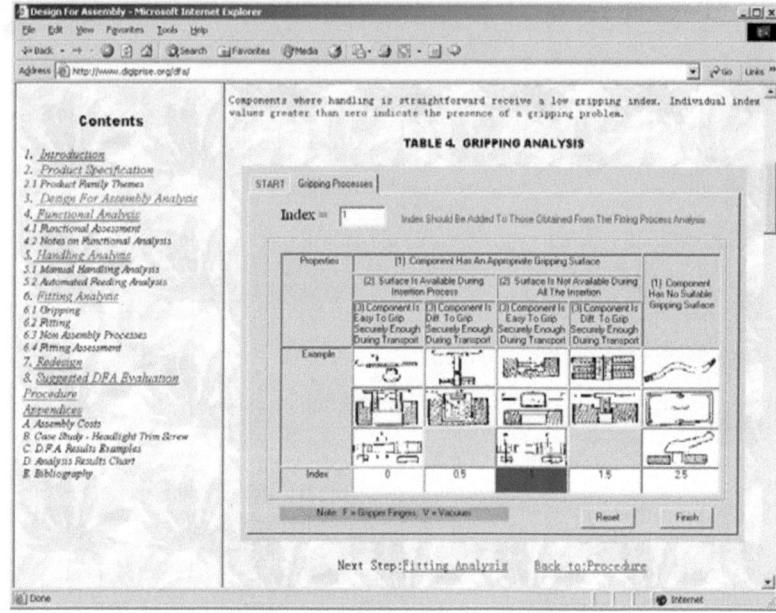

Figure 8.7 "Gripping Analysis" Web page.

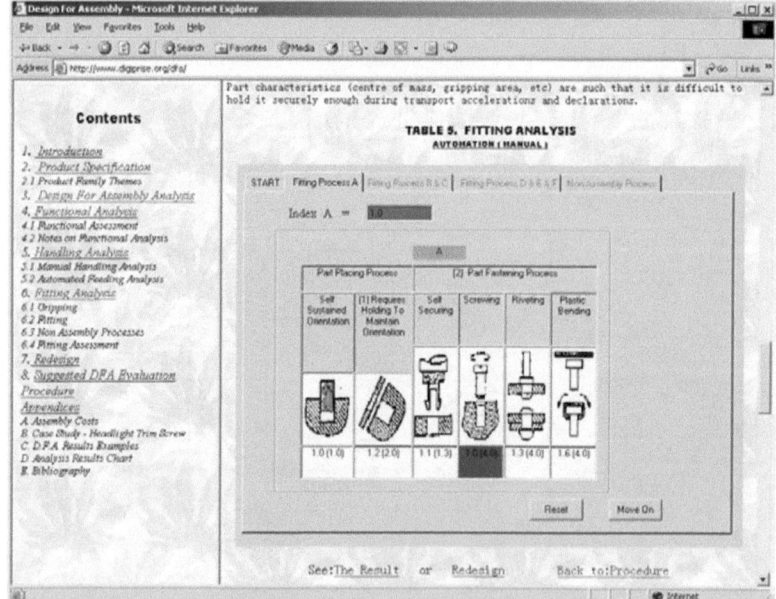

Figure 8.8 "Fitting Analysis" Web page.

8.2.3.   Web-Based Handling Analysis

The Lucas DFA distinguishes between manual handling analysis and automatic feeding analysis. Accordingly, two Web pages are created. Figure 8.5 shows the "Manual Handling Analysis" Web page. Manual Handling Analysis is conducted to one part at a time and the results are shown in the first tab page. The data control used to switch among different parts is also in the first page. There are another three Tab pages: "Size and weight", "Handling difficulties", and "Orientation", corresponding to three major factors affecting assembly handling. These pages are active. Clicking over one area triggers further system actions such as recording the corresponding data in the worksheet. Similarly, "Automatic Feeding Analysis" Web page is created, as shown in Figure 8.6.

8.2.4.   Web-Based Fitting Analysis

The last step in the Lucas DFA is "Fitting Analysis". Fitting involves three basic tasks: gripping, insertion, and fixing. Non-assembly operations are also assessed during the Fitting Analysis. Fitting Analysis requires the specific sequence of assembling the product. The Lucas DFA suggests that the assembly flowchart should be conveniently used to model the assembly fitting operations. One feature of the assembly flowchart is that assembly operations are attached to product parts. Therefore, the flowchart is regarded as part of the worksheet. It is envisaged that an assembly flowchart editor should be incorporated into the worksheet.

Two web pages have been created. One shown in Figure 8.7 is dedicated to support Gripping Analysis. The other shown in Figure 8.8 is to support Fitting and Non-assembly Processes. Both pages are made of the multi-page tab. The first tab page, as usual, is for recording the intermediate results. The data control used to switch among different parts is also in the first tab page. The remaining five, one in Gripping page, four in Fitting page, are correspond to the types of tasks involved in the fitting process. If an operation is of the gripping type, then a gripping analysis is carried out using the "Gripping Analysis" Tab page. If an operation is of the insertion type, then an Insertion Analysis is carried out using the "Insertion Analysis" Tab page of the "Fitting Analysis" Web page. If the operation is of the non-assembly type which includes fixing, then a corresponding analysis is carried out using the "Non-Assembly Processes" Tab page.

8.2.5.   Collecting Results

Results from Functional Analysis, Handling Analysis and Fittings Analysis can be collected on the same "Results" Web page. In the result web page, an ActiveX control of MsFlexDBGrid is used to show the analysis results of all of parts. Above it, some DFA indexes are presented. The analyst would formulate solutions based on them.

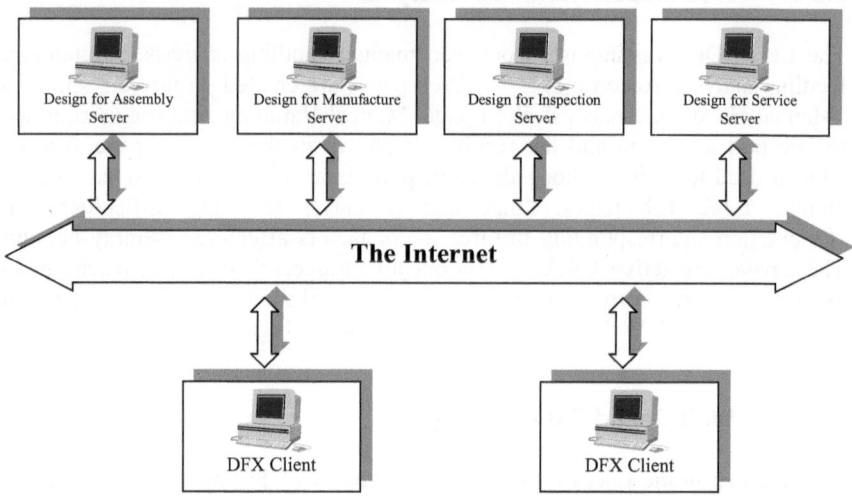

Figure 8.9 Client-server web-based architecture for collaborative DFX.

Table 8.1 Major activities involved in the web-based DFX analysis cycle

Activity	Where?
Start HTML browser	User
Start HyperLink to connect with the DFX server	Client
The DFX server sends its first HTML web home page	Server
Follow the first home page to start DFX analysis	User
The user enters inputs as appropriate and necessary	User
Carry out data integrity check and minor processing locally	Client
Submit the data to the DFX server	Client
Receive data from the client	Server
Convert data appropriately by executing a CGI program	Server
Call the application program to access the DFX data/knowledge base	Server
Convert the out data from the application program into HTML page	Server
Send the output HTML web page to the client	Server
Receive and render a new output HTML page from the server	Server
Repeat Step 4 until the DFX analysis is completed	Client

8.3.   IMPLICATIONS FOR FUTURE DEVELOPMENTS

The above section shows that the current web technology can be exploited to provide DFA technique in particular and DFX tools in general on the Internet. This simple experiment has wider implications. This section discusses four of them. They are (1) client and server architecture for collaborative product development; (2) Generic framework for developing DFX tools; (3) integrated utilisation of multiple DFX tools; and (4) Integration with other decision-support systems such as CAD, CAPP, CAPM in the product development process.

8.3.1.   Web-Based Architecture for Collaborative DFMA

The Internet is practically a synonym for client and server computing. The information is stored on servers, and clients make requests for the information they need. The server and client machines use the Hyper Text Transfer Protocol (HTTP) to exchange messages in the HTML format. Both HTTP and HTML are open standards and are implemented on a wide variety of platforms on the Internet. Since the data can be plain documents, sounds, or images, proper viewers need to be launched on the client side. Microsoft Internet Explorer and Netscape Navigator are two most widely used browsers. The browser, which is the software that runs on the client, requests a page from the server. The server sends an HTML file that describes the page. Once the file arrives at the client's site, the browser must reconstruct the document and display it on the screen according to HTML description it received from the server. A web server's job is to supply the HTML files requested by clients. A browser's task is to render the HTML document on the client's screen.

With the web-based client and server architecture, international centres of excellence can be networked through the Internet to provide the best services to the client users. Assume that each centre is an expert at one aspect such as Design for Assembly, Design for Machining, Design for Inspection, Design for Service, etc. Expertise and knowledge is incorporated into a web-based computer software package at a certain location on the Internet. These computers are DFX servers where DFX expertise resides. A project team has the access to these services through client machines connected to the Internet. The user uses an appropriate HTML browser to connect to and subscribe services from a DFX server. The browsers help users form a request, send it to a server, and present the users with the results from the server. A server receives and validates the request, retrieves data, and delivers them to the requesting client. Figure 8.9 shows a conceptual overview and the general procedure is explained in Table 1.

When the web-based client and server architecture is applied to collaborative Design for Manufacture and Assembly, a number of issues must be resolved. For example, how computation is shared between the server and the client, and how a high level of interactivity is guaranteed at the client site.

Computation is distributed between the server and the client. Generally speaking, servers are powerful machines and therefore should be used for tasks involving more number crunching. On the other hand, local clients are less powerful but with good Internet connectivity and therefore should be mainly for checking input and output constraints.

If all the computation is carried out by the server, then the client must submit every bit of the user input to the server. Clients and servers communicate to each other through HTTP by exchanging HTML files. This process is in many cases more time consuming than data processing by either the server or the client. Because of this limitation, it becomes problematic to maintain a reasonable level of interactivity between the client and the user. That is, the user often has to wait for the client to contact the server and receive HTML files. This is particularly undesirable in DFX analysis which is extremely interactive. That is, once the user inputs a piece of information, the DFX system must react instantly. This is at least how well-known DFA systems work.

In order to overcome the limited interactivity, client-side scripting and processing is necessary. In fact, the experiment reported in Section 8.2 has been done solely on the client side. DFA data is included in the program scripts. Therefore, the resulting system is as interactive as any other computer aided DFX system.

8.3.2. Web-Based Generic Framework for Developing DFX Techniques

a and Mak (1997) have proposed a generic DFX framework for developing and applying a variety of DFX techniques. This framework considers the product realisation process as a triple (P, A, R) of Products which compete in the market, Activities which realise products, and Resources which are available for realisation. P, A, and R are interrelated to each other. Interactions can be explained that products consume activities and activities consume resources. X in DFX defines the focus of a DFX tool.

Huang and Mak (1997b, 1998) use the framework to show a seven-steps procedure for developing DFX techniques. They propose to use a number of formal but pragmatic "common-sense" constructs to convert the conceptual DFX framework into a practical platform. For example, bills of materials are used to describe and analyse the overall product structure and product characteristics. Flow process charts are used to describe and analyse the overall process structure and process characteristics in relation to individual product elements. Standard operation process charts are modified to describe and analyse the overall process structure in relation to the product structure. Appropriate performance measures are used to evaluate the interactions between the elements of products, processes, and resources.

At present, the following code components are being investigated for possible deployment on either the client side or the server side. These code components are briefly summarised as follows:

- "Product Analysis" component. Every DFX technique requires product data to assess design decisions. A generic code component can be incorporated into the web-based DFX system. It is basically a Bills of Materials (BOM) editor, possibly supported by a facility for capturing key characteristics of product items.
- "Process Analysis" component. Every DFX technique is concerned with a particular business process from market research through design and manufacture to product retirement. A generic component is needed for the user to systematically capture process characteristics necessary for the analysis. This component is basically a flow chart editor, possibly supported by a facility for capturing key characteristics of process activities.
- "Interactions" component. Basic to most DFX techniques is the examination of interactions between product elements and process activities. Operation/route process charts can be modified for developing a generic "Interactions" component, just like the "Assembly Flow Chart" used in the Lucas DFA technique. This component is basically an operation flowchart editor.
- "Worksheets" component. Reporting facilities are an important feature of most successful DFX techniques. They reflect the logical workflow of activities involved during the analysis and keep track of intermediate results. For example, interactions between products and processes are first established. Their mutual consumption is then measured and compared with the benchmarks. Strengths and weaknesses of the current product and process design are highlighted. Causes are diagnosed and redesign advice is formulated. This component can be easily implemented by standard multi-page and grid components.
- "DFX Handbook" component. Most systematic DFX techniques are based on comprehensive DFX data and knowledge bases, often called DFX manual or handbook especially in paper-based form. Checklists and look-up tables can be used to compile such handbooks. There are two modes in which such controls should work. One is the compiling mode for DFX handbook, whether in the form of guideline or lookup tables. The other is the application mode to extract relevant data from the DFX handbook.

The first three components are relatively domain independent. That is, they can be shared, with little modification, by various DFX techniques. They provide common platforms for integrating and collating results from individual DFX analyses. On the other hand, all the domain-specific elements can be encapsulated within the "Handbooks" component. For each web-based DFX technique, a "Handbooks" component must be designed and developed individually to include data and knowledge specific to the chosen business process.

8.3.3.   Web-Based Framework for Integrated Utilisation of Multiple DFX Tools

The use of a single DFX technique may stretch the product design in one direction and upset the other aspects. It is therefore highly desirable to employ multiple DFX techniques to achieve overall optimum solution. This section presents a scenario where multiple DFX servers are available on the Internet for the same product development project. There are two levels of integration: tight and loose integration.

If all the individual DFX tools are developed following the generic framework mentioned previously, then a tight integration of these tools is very much facilitated. Even if they are distributed on the Internet, it is still straightforward to integrate them through careful client-side scripting. The scripts can be packaged into an ActiveX code component so that it can be used on any web-based DFX pages. Figure 8.10 shows a web page for tight integration of DFX tools. A drop-down Combo box is used to switch between different DFX tools involved one at a time. The main component is the facility consisting of multiple tab pages. These pages correspond to major activities in the DFX analysis as described in Huang and Mak (1997b). They include product analysis, process analysis, interaction analysis, performance measurement, etc. In addition, there are other tab pages for trade-off analysis since multiple DFX tools are involved. Trade-off analysis is mainly concerned with collating results from individual DFX tools and proposing overall solutions.

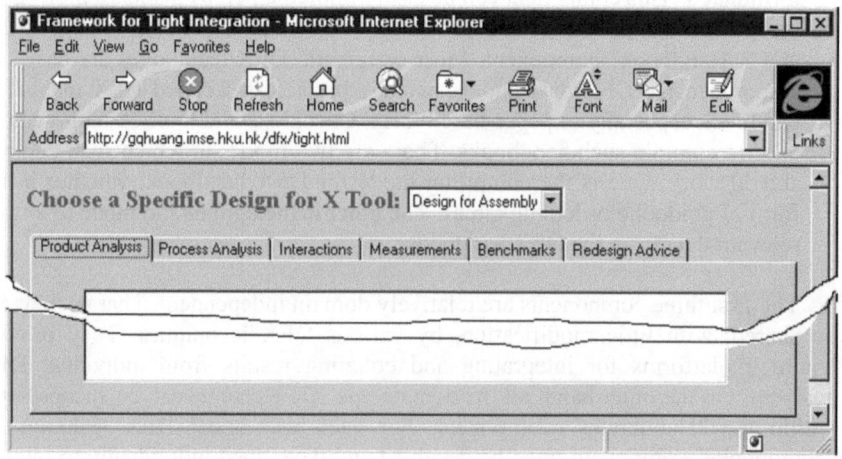

Figure 8.10 Web-based framework for tightly integrated use of multiple DFX tools.

In reality, it is extreme rare that two DFX tools follow the same framework, not even those developed by the same person or group. In fact, there are wide differences between existing DFX tools in terms of product and process data representations, and DFX data/knowledge representations. A tight integration as described above is not usually feasible. Instead, only a loose integration can be achieved. Figure 8.10 shows a web page for loose integration. Once again, the multi-page facility is used. The first tab page provides hyperlinks to individual DFX servers in a logical way so that they can be activated according to the need of the project. For example, Boothroyd (1996) suggests that DFA should be conducted first to ensure an overall simplicity and then other types of DFX technique follow. The rest tab pages are created to keep results from individual DFX tools. Although a tab page can be created to collate the overall results from participating DFX tools, it is the analyst who integrates and considers the individual outcomes and works out overall solutions.

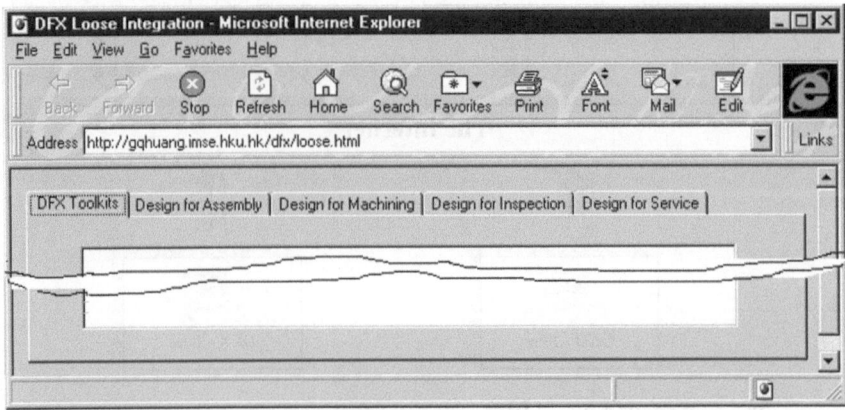

Figure 8.11 Web-based framework for loosely integrated use of multiple DFX tools.

8.3.4.   Web-Based DFX-Oriented Collaborative Product Development

The integration of various decision-support systems for collaborative product development has long been the concern of many researchers and industrialists. Leading research groups have resorted to the web-based Internet technology for possible solution. A general discussion of web-based collaborative product development is beyond the scope of our research. Instead, this chapter is particularly interested in enabling collaboration in product development by using web-based DFX tools.

The essence of DFX is concerned with the interactions between product characteristics, process activities and plant resources. DFX analysis is an important activity that integrates other decision-making activities in product development. This can be described by a scenario of DFX-oriented collaborative product development. There are two ways for DFX to acquire necessary data. One is to acquire (1) product characteristics from a CAD system; (2) process activities from a CAPP system; and (3) plant resources from a CAPM system. This would involve direct communication between these application servers.

In the other approach, DFX acquires necessary information from a web-based Product Data Management (PDM) server, not directly from individual application servers. That is, product, process and resource data are maintained in a Product Data Management (PDM) server This approach is consistent with those proposed by Kim et al (1996).

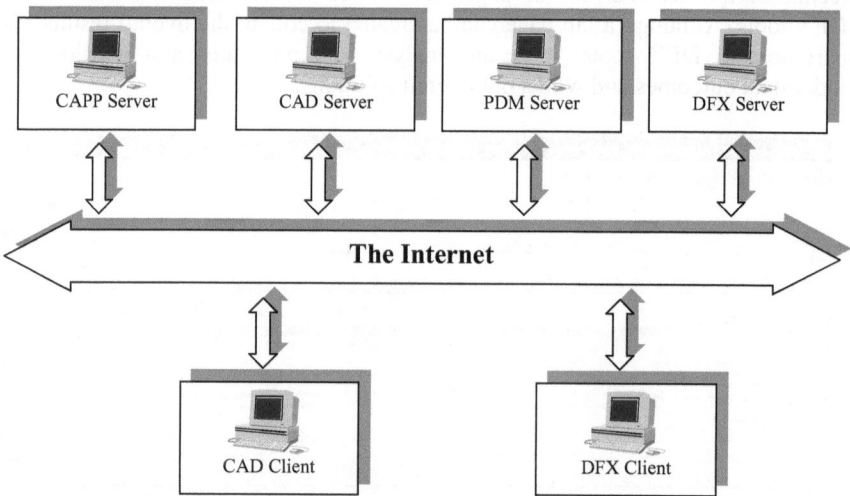

Figure 8.12 Web-based DFX-oriented collaborative product development.

Figure 8.12 shows that application servers and clients are connected through the Internet. Application systems send and acquire appropriate data to and from the PDM system. The PDM server plays an important role here. It not only maintains product, process and resource data, but also provides a collection of translation services from one format to another through the CGI programs.

Let us first consider the process of storing information into the PDM database. A CAD application client uses the corresponding web-based CAD server to design a product or a part. At the end, resulting data are submitted to the PDM server which uses a CGI program to translate the data in HTML format into a required format and stores the converted data in the database. Other web-based decision-support systems work in a similar way.

Let us now look at how DFX acquires information from the PDM server. The DFX browser sends a message to the PDM server. Upon the receipt of the request, the PDM server invokes CGI (Common Gateway Interface) program which interprets the request string into the appropriate format for the server to retrieve relevant product data. Another CGI program is invoked to convert the product data from the PDM format to a format suitable for HTML browser. Process and resource data can be acquired in the same way from the PDM server.

The results from DFX analysis are also stored in PDM server. For example, the "Result and Report" web page of the DFX submits the outcome to the PDM server where an appropriate CGI is executed to convert the data into a suitable format for storage. This process is similar to the way that other applications programs such as CAD, CAPP and CAPM ca submit the results to the PDM server.

8.4.   SUMMARY

This chapter has demonstrated how DFX techniques can be made available on the Internet. The experience with the experimental web-based DFA has advantageous implications. The recent development in web-based technology makes it possible to implement web-based DFX techniques functionally equivalent to those traditional computer aids on standalone workstations. Furthermore, there are wide choices of third-party ActiveX controls. They can be readily exploited and modified for implementing DFX components, including product analysis, process analysis, impact analysis, and data display and presentation, etc. These features significantly speed up the DFX development process and facilitate the application as well.

Installation and maintenance are no longer necessary on the client side. As long as the user has the use of a web browser, he or she can have instant accesses to any web-based DFX technique available on the Internet. The DFX client collects product, process and resource data and the interacting relationships, and submits the data to the DFX server. Upon receipt of the necessary data, the DFX server first interprets the input data into a format which can be used for searching the assessments from the DFX handbook (database). The resulting data are then translated into HTML format to create a web page for the DFX client browser. The user can take further actions according to the outcome shown on the returned page.

The client-server architecture is a metaphor appropriate for collaborative product development, especially when the tools or teams are geographically distributed. Both the client and the server communicate to each other using a standard HTTP (HyperText Transfer Protocol), regardless of their hardware configurations and operating systems. Two integration scenarios have been discussed in this chapter. One is the integration of multiple web-based DFX tools through the Internet. The other scenario is the integration between other web-based decision support systems such as CAD, CAPP, and CAPM in the product development process.

Our future work would focus on the development of a number of generic facilities in the form of ActiveX components for rapid customisation of web-based DFX techniques on the Internet and the development of a web-based Product Data Management (PDM) system on the Internet for the integrated application of multiple DFX tools.

This chapter has demonstrated how Lucas DFA techniques can be made available on the Internet. The experience of the experimental Web-based DFA has shown that it is possible to implement web-based DFA techniques functionally equivalent to those traditional computer aids on standalone workstations. Furthermore, Web-based system has some advantages, for example, analysis system installation and maintenance are no longer necessary on the client side. As long as the user has a Web browser, he or she can have instant accesses to any Web-based DFA technique available on the Internet. The DFA client collects product, process and resource data and the interacting relationships. The web server will exploit and modify the system for client users, including product analysis, process analysis and data display and presentation, etc.

9

FAILURE MODE AND EFFECT ANALYSIS (FMEA) OVER THE WWW

Over the last two decades or so, good design practices have been formalised into a suite of techniques and methods – design tools. Some examples include Quality Function Deployment (QFD), Functional Analysis (as used in Value Analysis), Failure Mode and Effect Analysis (FMEA), Fault Tree Analysis (FTA), Design for Manufacture and Assembly (DFMA), and Morphological Chart Analysis. This chapter reports on another early attempt of the authors on using FMEA as an example to demonstrate how such design tools can be made available on the Internet through the web browsers. The purpose was to gain some insights concerning the strengths and weaknesses of the various approaches to the implementation of web-based design tools. Indeed, experiments have been carried out on at least three versions of the web-enabled or web-based FMEA.

This chapter is intended for practitioners who are considering the conversion of their design methods and techniques into web-based virtual consultants. Section 9.1 gives an overview of the roles and challenges of formal methods and techniques in product design and manufacture. Section 9.2 presents an overview of the web-based FMEA system in terms of its architecture and main components. Issues related to system development, implementation, operation and evaluation are reported in Sections 9.3 and 9.4 respectively. Finally, some suggestions are specified in Section 9.5 regarding various implementation techniques.

9.1.   DESIGN TOOLS AS VIRTUAL CONSULTANTS ON WWW

In recent years, design tools have enjoyed not only significant attention from academic researchers, but have also been a topic of interest among industrial practitioners. Most of the design tools originated actually from industries to deal with specific industrial problems. The progress and the current status can be seen from recent industrial surveys. Wright et al (1996) report on the use of formal design tools in the UK manufacturing industry. Norell (1993) reports on the industrial practice of some design tools in Scandinavian manufacturing industry. McQuater and Dale et al (1996) survey on the use of formal design tools as quality management tools in five manufacturing plants. Industrial surveys have also been conducted on individual design tools. For example, Dale and Shaw (1990) have carried out a survey of FMEA. Pandey and Clausing (1991) have conducted a survey of QFD.

These formal design tools have been increasingly applied in manufacturing industries. Indeed, FMEA has been widely standardized, MIL-STD-1629A in the USA and BS 5760 in the UK. The industrial users have reported significant benefits of these design tools. Successful users have achieved typically 15-45% in quality improvement, costs and time-to-market reduction. Above all, intangible benefits also prevail. Their use not only rationalises the products and the associated processes but also rationalises the product development process itself. Most of such tools are team tools, and they should be used as according to their nature. An extra benefit is their ability to build team spirits and to facilitate teamwork.

Despite the above achievements, however, disappointments have also been expressed by academics in that design tools have not yet been practised to an extent to have their potentials maximized (Andreasen, 1987). Gill (1990) has given some of the reasons for this. One reason is that formal design tools are not familiar to industrial practitioners, because these tools were not taught widely at colleges and universities in the past. In fact, most practitioners generally acquire knowledge of the design tools through short course training. Such post-experience training is not as effective as expected, because engineers have already developed their own ways of doing things, and it is very difficult to change their habits to follow the procedures of formal design tools. In addition, most formal design tools often require intensive data inputs from a number of disciplines. Teamwork thus becomes a must to achieve such data requirements. To put it in another way, the use of formal design tools demands and thereby promotes teamwork. Besides, most design tools involve a fair amount of paperwork and, therefore, may be time-consuming to use. Gill has pointed out a number of difficulties in implementing formal design tools. By using an in-depth industrial case study, he demonstrates (1994) that, if the inquiry is at the level of philosophical foundation, basic tenets, and value systems, the causal relationships between design tools and successful enterprise can be much more significant than were initially expected.

One way of overcoming these difficulties is to computerise these design tools. Computerisation is viable because they are usually packaged into pro-forma worksheets together with a systematic procedure of instructions. Reinders (1995) discusses a number of computer based design tools, or design assistants, as he names them. Numerous computerised FMEA systems are commercially available (Technicomp, 1990; http://www.fmeca.com). Computerised design tools do not completely remove these difficulties, and in fact introduce some other overheads, such as familiarity with operating a computer software package. While recent development in multimedia computing fortunately enables the ease of use, most computerised design tools are packaged as standalone systems. Standalone design tools support teamwork in a limited way. This phenomenon can be described as isolated "Islands of Expertise". It is not consistent with their tenet of facilitating communication, fostering teamwork, and collaborative product development environment.

This chapter proposes to resolve the issue of "Islands of Expertise" by exploiting the rapidly developing Internet/Intranet technology. Emphasis is placed on a specific design tool – FMEA. Failure Mode and Effect Analysis (FMEA) is one of the formal techniques for effective product development. Its main purpose is to avoid as many potential failures as possible by identifying them and taking appropriate actions in the early stages of design and development. This paper proposes to employ the World Wide Web (WWW, web) technology to provide FMEA services on the Internet/Intranets. Its outputs, the web-based FMEA systems, require no installation or maintenance but offer remote and simultaneous accesses and therefore better teamwork.

9.2. SYSTEM ARCHITECTURE OF WEB-BASED FMEA

The prototype web-based FMEA system consists of three main components, namely, the FMEA web server, FMEA database server, and FMEA clients. These components may be located in different geographical sites, but all are linked through the Internet. For example, the FMEA web server may be located within the research group at the University of Hong Kong, whereas the FMEA database server is located in the company's Hong Kong headquarters, or in a manufacturing plant that is more proximate to potential clients for quick access. Clients can be anywhere in disparate regions, as long as accesses to web browsers on the Internet are available to them. In this connection, the three main components are served by three different computers. Each computer has a unique IP address for identification during access. Figure 9.1 gives an overview of setting up the 3 components.

Microsoft Internet Information System (IIS) acts as WWW Server, and provides the HTML files and the CAB files that are needed by the FMEA system. A web site is created for the FMEA system at this web server that actually includes web sites for other applications. It receives the client-side request, processes data and, if necessary, communicates with the FMEA database server.

The FMEA web site includes a number of web pages. The FMEA home page includes hyperlinks to other FMEA functional pages. The FMEA worksheet is the main web page. Users can use these pages to enter or retrieve FMEA data, and to carry out a variety of analyses.

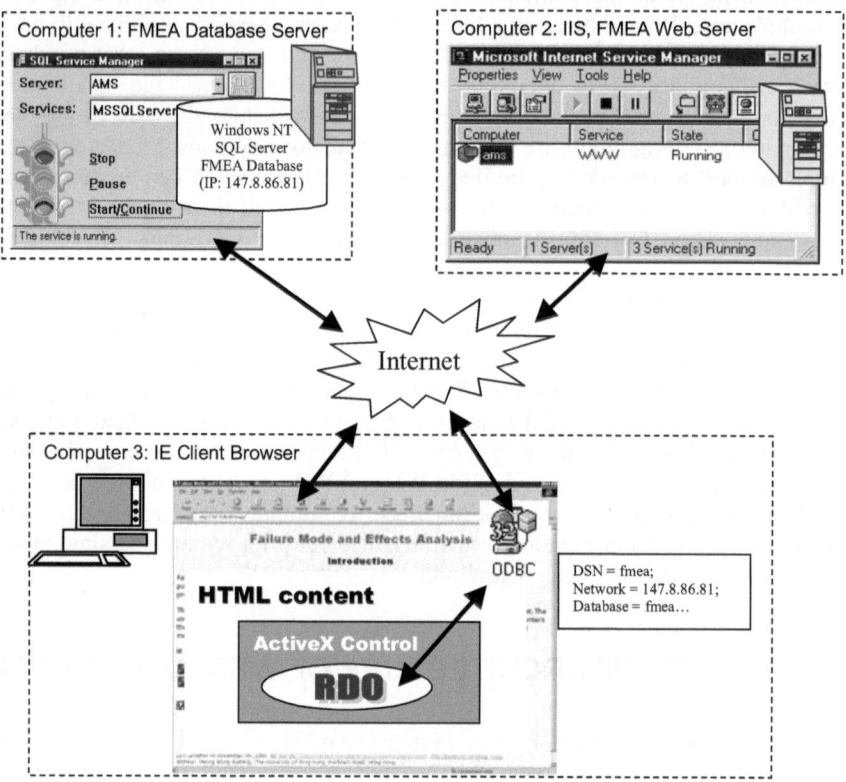

Figure 9.1 Web-based FMEA architecture.

The relevant FMEA data are managed by an FMEA database server that uses the Microsoft SQL server - a relational database management system with a client-server architecture. The FMEA database server and the FMEA web server may be physically implemented on different computers, as experimented in this work. Thus, this FMEA database is remote as far as the FMEA web server and the clients are concerned. Access to this remote FMEA database can be achieved from the FMEA server or the clients, depending on how the FMEA is developed and deployed. Both may use the ODBC (Open Database Connectivity) technology. In the current implementation, the FMEA system actually accesses the remote database through the client. The client must create an ODBC data source on the client machine. The data source is then used to obtain and transport data items from and to the remote database through Remote Data Object (RDO) and controls

(RDC). Since the data source is created on the client machine, the SQL ODBC driver must also be made available on this client machine. This requirement cannot usually be met in most clients, and is indeed a serious shortcoming of this approach to remote data access.

FMEA clients are not actually part of the web-based FMEA system, and are totally unconnected to FMEA prior to connecting to FMEA. They become part of the system when they are connected to the FMEA web server. In theory, an unlimited number of FMEA clients can be using the services provided by the FMEA web server simultaneously, and the FMEA system may be used by different clients within the same project or in different projects.

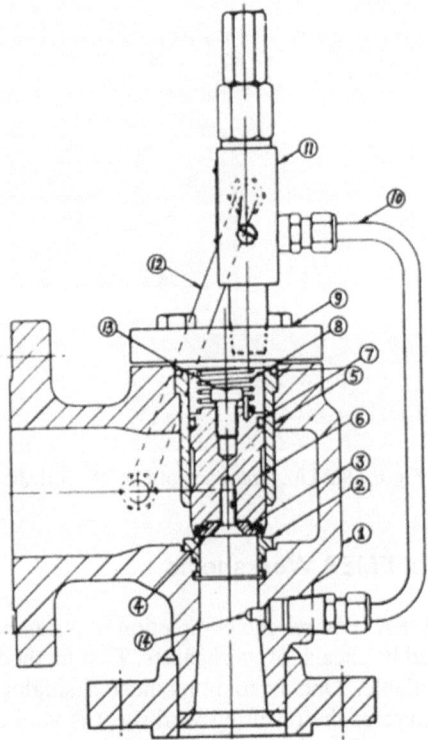

Figure 9.2 Sample product used for prototype development and preliminary evaluation.

9.3. FMEA SYSTEM DEVELOPMENT

There are wide variations in FMEA. In terms of formats, FMEA may take the form of matrix, table, and chart. As regards contents, different items and information are included. For example, BS5760 and MIL-STD-1629A are slightly different, but both are very different from the Ford Motor Company FMEA that has been selected for the present study. A sample product shown in Figure 9.2 is used to build up the database and to test the program codes incrementally.

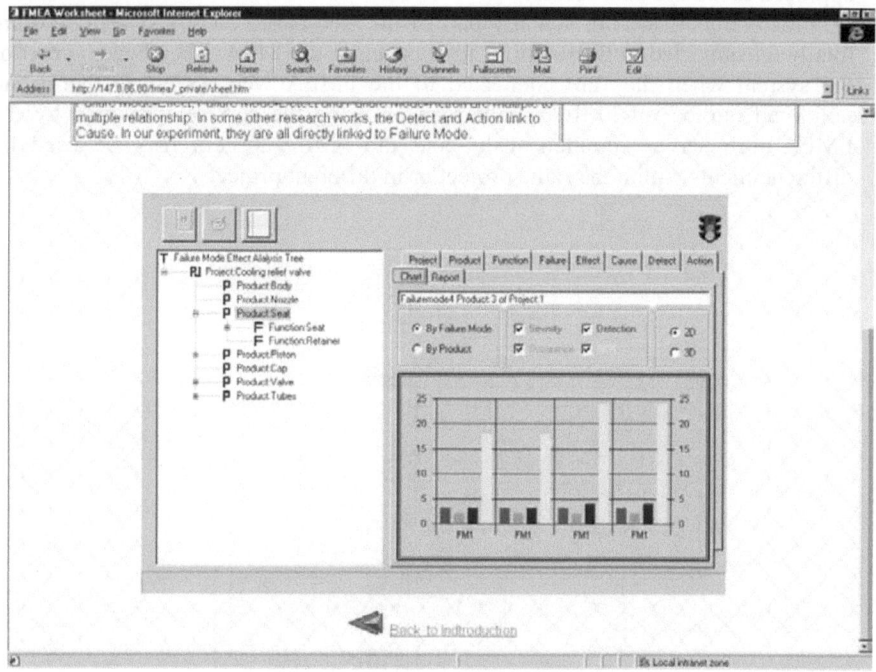

Figure 9.3 User interface of FMEA worksheets.

9.3.1.   Design of FMEA Worksheets

The operation of FMEA requires the completion of a number of activities. This is usually accomplished by means of worksheets. Two methods of implementing the worksheets are available. One is to implement a single worksheet for all the activities. The structure of the resulting worksheet is very complicated, including all the details on a single form just as the paper-based FMEA. It is very difficult to fit all these details in the limited space of the screen. The other choice is to implement one form or screen for one FMEA activity. For example, one form is used to define product items, one form is to define failure modes, and so on. An advantage of this approach is that the user can always have a very specific focus at any time, and the downloading time for each form may be short. The disadvantage, however, is that it is difficult to present the user with an overall picture of the analysis.

In the present study, a compromising approach is adopted. Figure 9.3 shows the main user interface of the worksheets, and there are two main areas on the screen. One is the hierarchical tree structure on the left-hand side, and the other is the component of the multiple tab pages on the right-hand side. The tree diagram on the left-hand side depicts the structure of the subject items (products, processes, or systems) and an analysis of its failures. Whereas each tab page on the right-hand side is also dedicated to one activity or one group of closely related activities, so that the details of each activity can be defined and displayed. The functions of the tab pages are briefly summarised as follows:

- The "Project" tab page. This involves the specification of the heading details, such as project title, responsible person, person in charge, date of analysis, and so on. It is also suitable to specify the remote FMEA database. Information is used to define an ODBC on the client or the server, depending on how the worksheet is deployed.
- The "Product" and "Function" tab pages. In editing the tree diagram of items, this page provides facilities to define the details of the subject product or process or system, etc. Although the "Function" tab page is provided separately, it is considered more suitable to combine them to form an "Item" tab page in the future. This is because some FMEA is conducted to items without involving their functions.
- The "Failures" tab page. This page lists all the failure modes of the selected item, as highlighted in the tree diagram. The editing facilities are provided for adding new, modifying existing, or deleting obsolete failure modes. The severity (S), occurrence (O) and detection (D) are specified and the Risk Priority Number (RPN) is evaluated for each failure mode.
- The "Effects", "Causes", "Detect", and "Actions" tab pages. These pages are used to display and define effects, causes, detection methods, and actions of a selected failure mode, as highlighted in the "Failure" tab page or the tree view, respectively.
- The "Chart" and "Report" tab pages. The "Chart" page provides various facilities to display item failures according to different criteria. For example, the failures of a selected item can be displayed in a ranking order in terms of S, O, D or RPN. It becomes easier for the user to identify the most significant failures or items. The "Report" tab page provides the tabular display of the equivalent information in the "Chart" page. A final FMEA report can be produced as a separate web page.

9.3.2. System Implementation and Deployment

The FMEA system (worksheet) has been developed initially as a standalone program and, afterwards, deployed as an ActiveX control. The FMEA ActiveX control is then incorporated into an FMEA web page. The ActiveX technology is used for several reasons:

- There is no need to develop CGI programs.
- High interactivity between the user and the system can be achieved.

- Friendly user interface for the system.
- The absence of overhead communication allows the clients to perform complicated processing with the server.
- No installation or maintenance is required.

There are three basic ways of deploying the FMEA worksheet and remote data access on the Internet, namely, on the server side, on the client side, and distributed on both sides. Figure 9.4 shows the first two ways, i.e., that of the server and the client. When deployed on the server side, the client browser is only responsible for collecting input data from and displaying outputs to the user, as indicated by lines 1 and 4 in Figure 9.4, while computation is carried out by the server machine. This method, however, is not suitable for applications where user/system interaction is frequent, because the communication overheads between the server and the client may become too expensive. In this deployment mode, the FMEA web server would be entirely responsible for communicating with the FMEA database server, which is illustrated by lines 2 and 3 in Figure 9.4.

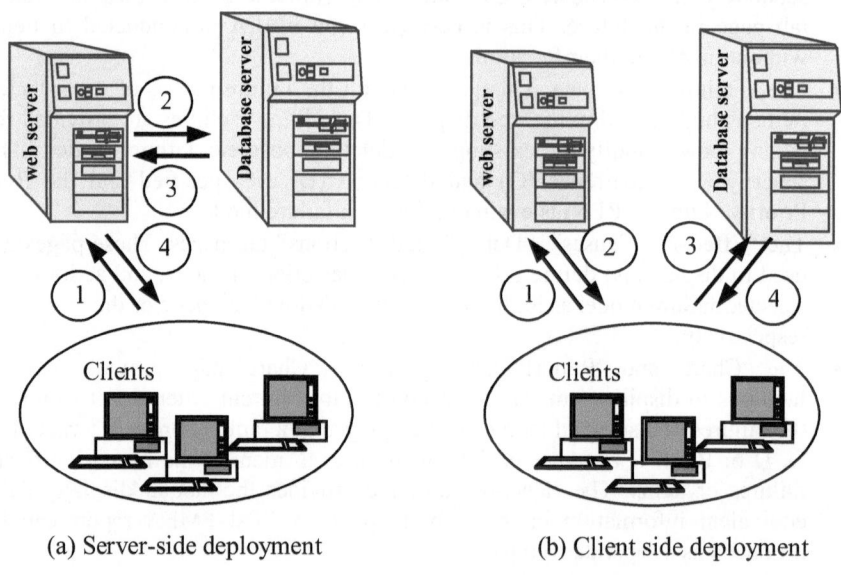

(a) Server-side deployment   (b) Client side deployment

Figure 9.4 Server-side versus Client-side deployment.

The other extreme is to deploy the FMEA worksheet on the client side. In fact, the FMEA control is embedded into a web page. The control and its hosting page are downloaded onto the client machine when the user connects to the server page (lines 1 and 2 in Figure 9.4). Computation takes place at the client machine. No further communications are necessary with the server. Although its downloading time may be long, its most significant advantage is that the user can interact with the system as frequently as necessary with no overheads incurring at communication with the server. Lines 3 and 4 show that remote FMEA database servers must be established by all the clients. Thus, the user has the freedom or is required to specify a remote data source based on ODBC. This requirement is not usually met in practice, because the SQL Server ODBC drivers may not be available on all the client machines. This flexibility advantage turns itself into a serious limitation.

One solution to compromise the above two extreme deployments is to distribute the FMEA system on both the FMEA web server and the clients. For example, the remote data access is deployed on the server side, while the worksheet is deployed on the client side. This type of distributed deployment provides the required high interactivity between the user and the client machine, while eliminating the need for each client to define an ODBC data source. In this type of deployment, the data source is created on the FMEA web server machine, where the SQL ODBC drivers are installed. The resulting data source is then used to obtain and to set data items to the remote database through ActiveX Data Object (ADO) and controls (ADC). The resulting set is then used to populate the contents of the ActiveX control of the FMEA worksheets that are deployed on the client side.

The appropriateness of each of the aforementioned applications is largely determined by the nature of individual applications. Although the distributed deployment is now considered more suitable for deploying the FMEA system, it has actually deployed on the client side at present. A major reason is that it is deemed more desirable to wait until such technologies as ActiveX Data Object (ADO) and control (ADC) become available in the chosen programming environment. The distributed deployment will be investigated as soon as the environment is upgraded in the very near future. Nevertheless, the FMEA report page has adopted the server-side deployment using the ASP (Active Server Pages) and ADO.

9.4.   SYSTEM OPERATION AND EVALUATION

The prototyped FMEA system has been developed and deployed as designed. This section describes the general procedure of using this system and discusses some of the preliminary findings obtained in the initial evaluation of the system.

9.4.1.   System Operation

Based on the existing design, implementation and deployment of the FMEA system, there are four main stages in its operation. First, Internet Explorer is used to connect to the FMEA web server. Undoubtedly, consideration has to be given to security issues regarding the access of information at its different levels of confidentiality, the responsibility of such rests on the authorities concerned. The FMEA worksheet is actually downloaded from the FMEA web server to the client machine during the access. The system is executed on the client machine for its full functions, having nothing to do with the FMEA web server any more. Before starting an FMEA, an ODBC data source must be created for the client for accessing the remote database. This can be done by using the registration facility in the FMEA worksheet. The first stage finishes with the definition of the details of a project.

The second stage is to define items (products and their functions to design FMEA and operations to process FMEA). Items are arranged hierarchically using a TreeView ActiveX control. Facilities such as mouse clicking, drag and drop, etc., are provided for the user to edit and browse the details and relationships of items. For example, a single left click on a function node in the tree retrieves its description and its associated failure modes. These data are then displayed and stored in relevant tab pages. Clicking the mouse on the right gives a list of commands to the user to choose. For example, if the selected node is a product node, the user may choose one command from the list of "Update Product", "Add Product", "Delete Product", and so on.

The third stage is to define the failures of the subject items one after another. The user selects an item either by clicking on the corresponding tree view or using the drop-down box. The existing failure modes of the chosen items are displayed. There are available built-in and user-friendly facilities for the user to update, add and delete failure modes, including their severity (S), detection (D), and occurrence (O). Their Risk Priority Numbers (RPN) are assessed once S, D and O are fully specified. The failure definition also includes identification of the causes, effects, detection methods, and the preventive/corrective actions of the chosen failure mode.

The last stage is to analyse and report the results. Charting facilities are provided for the user to rank in order the items and/or failure modes in terms of severity, occurrence, detection, and RPN. This helps the user to identify the most influential items and failures from various points of view, and even the minor ones will be given the necessary attention. A number of reporting facilities would be made available. At present, the system generates a final report in the HTML format on the browser. The user may print hard copies of the final reports for their own references.

9.4.2.   System Evaluation

Preliminary tests have been conducted within limited scope. Initial findings are obtained regarding the existing design, implementation, and deployment. In terms of the capability, the prototype system performs functions as designed, and performance is well within expectation. In general, all the intended FMEA functions, as provided in standalone FMEA software systems, are demonstrated satisfactorily.

In term of performance, the prototype system is considered acceptable within the test environment in terms of the time required for downloading the FMEA worksheets, connection to the remote database, and working compatibility with the system after it is downloaded. However, proof has yet to be obtained on whether similar acceptable performance can be achieved if the FMEA web server, FMEA database server, and the clients are distributed far away. Problems have been encountered in accessing the remote FMEA database. This is mainly due to the fact that a data source must be created at the client machine. This problem is expected to be eliminated when the server-side deployment of the data access is introduced.

The prototype system has yet to be tested on its usability within industrial environments. Due to the web's multimedia capability, the user interface of the web-based FMEA can be developed as good as, if not better than, that of standalone FMEA. In this connection, it can be reasonably inferred that, if the usability of standalone FMEA is acceptable to industrial practitioners, the usability of the web-based FMEA should also be within acceptable limits.

In view of the chosen programming and deployment environment and ActiveX technology, the resulting systems are only supported by MS Internet Explorer at present, and improvements have yet to be made to enable it to be supported by other web browsers. This is inconsistent with the authors' original intention to adopting open standard web browsers. Although this can be overcome by converting the VB programs into Java programs, this will not be carried out in the near future during the prototyping period. In addition, the security level of the web browser must be customized, since the FMEA ActiveX control is not electronically signed.

9.5.   IMPLEMENTING AND DEPLOYING THROUGH THIN CLIENTS

In Section 9.3, it was discussed that the main FMEA component was first developed as an ActiveX component and then deployed (embedded) in a web page. The database connection was also based initially on RDO (remote data object) and later, on ADO (activeX data object). This is a typical fat-client implementation, thereby inheriting the main characteristics of this approach (What approach?). For example, one limitation is that the downloading time is relatively longer (could be considerably long depending on the communication channel). A notable strength, however, is that once downloaded successfully, its working performance is very fast and interactivity is excellent.

Some minor technical hitches were encountered by the authors in deploying the FMEA component. One concerns the need on the user's side to adjust the security settings of his/her web browser to enable the execution of ActiveX components. Another snag was that the ActiveX components are simply downloaded without being opened/executed within the web browser if they are not properly deployed. Finally, the user is sometimes required to install some components or resources used in the ActiveX components on the client-side. This is extremely undesirable.

In order to avoid the above limitations, experiments have also been conducted with the thin-client approach using Active Server Pages (ASP). Results from such experimenting attempts reveal that this thin-client approach is sufficient to provide the level of interactivity required in design tools such as FMEA, while overcoming most limitations of the fat-client approach.

However, it is important to point out that the fat-client approach adopted in web-based Design for Assembly was an appropriate choice because of its requirement of high interactivity between the client and the user.

9.6.   SUMMARY

FMEA is a team tool. Standalone software FMEA systems, however, provide limited support to the teamwork required in carrying out an FMEA. This paper has discussed the design, development and the initial evaluation of a prototype web-based FMEA system on the Internet/Intranet. Web-based FMEA provides better support to teamwork. Different functions of diverse disciplines and geographical distributions can simultaneously make their individual contributions. Users may use different web browsers to connect to the FMEA web server all at the same time to contribute concurrently from different perspectives. For example, design engineers can use the system to establish the product structure and functions required in FMEA. Test and QC engineers can adopt the system to enter failures they have identified for a particular product and its parts and components. Product and process engineers may use the system to identify the causes and effects of specific failures. Managers may use the system to identify priority areas for allocating resources and responsibilities. As a result, better communication is achieved. Since this work is at its early stage, fine tuning will be required and other significant improvements will be attained with fuller tests subsequently. Indeed, more specific guidelines would be produced regarding the choice of the most suitable implementation methods for developing specific web-based design tools.

10

WEB-BASED ENGINEERING CHANGE MANAGEMENT

Engineering Changes (ECs) are a kind of modification in forms, fits, functions, materials, dimensions, etc., of products and constituent components. Indeed, the agility of an enterprise of today is best demonstrated by its ability to manage changes efficiently and effectively, and Engineering Change Management (ECM) poses a direct effect on the enterprise's product development process. However, ECs involve considerable complexity, and recent investigations in numerous manufacturing companies have revealed that the number of ECs active at any one time reaches a level that makes management by a paper-based system and by an ad hoc procedure incompatible. Although sophisticated computer aided systems with comprehensive functionality are available in the market, such systems have not been utilised to facilitate EC management (ECM) activities. Indeed, standalone computer aided systems are limited in supporting the multi-disciplinary teamwork in ECM, especially when they are distributed in terms of location and time.

This chapter aims at developing a web-based ECM framework to facilitate the sharing of information among various parties of disparate geographical locations and also to achieve simultaneous data access and processing. It is hoped that the proposed framework is capable of overcoming the time and geographical limitations in the paper-based and standalone ECM systems, thus enhancing considerably the effectiveness and the efficiency in managing ECs. The parties involved are able to use open standard web browsers to access such systems regardless of their locations and start times.

The work reported in this chapter is only one main part of a research project. The overall research methodology is as follows. Phase 1 examines the current ECM practices through a comprehensive literature review and a thorough industrial survey (Huang and Mak, 1997b; Huang and Mak, 1999a; Yee, Huang and Mak, 2000). This chapter also presents a review of the existing ECM technologies, such as computer aided Configuration Management and Product Data Management where ECM is usually a subset (Huang and Mak, 1997c). Phase 2 establishes a theoretical ECM framework to extract the good practice elements and to overcome the limitations identified from the review (Yee, Huang and Mak, 1998; Huang, Yee and Mak, 2000; Huang, Lo, Yee and Mak, 2000). Phase 3 is mainly concerned with the development of the web-based ECM platform that incorporates the theoretical model. The investigation is concluded by illustrative case studies for further study. This chapter is mainly concerned with phase 3 of the system development.

ECM is a complicated issue and has often impacted on other business activities/decisions involving decision support systems, such as CAD (Computer Aided Design), CAPP (Computer Aided Process Planning), PDM (Product Data Management), and ERP (Enterprise Resource Planning). This chapter only focuses on ECM activities to bring out a more thorough discussion.

This chapter contains six sections altogether, including the introduction and conclusion. Section 10.1 reviews the ECM literature to give a brief overview of ECM and the web technology. Section 10.2 proposes the web-based ECM framework. The main components in the framework are described. A detailed discussion on deploying and implementing the proposed in the web-based environment is presented at Section 10.3. Section 10.4 describes the main facilities provided by the system for managing ECs. Lastly, the conclusion summarises the key issues and highlights the main directions for future work.

10.1. LITERATURE REVIEW

A comprehensive literature review has been conducted on ECM and is still ongoing. A number of major CD-ROM databases in the libraries were searched for relevant publications. It has been noted, surprisingly, that the engineering change (EC) literature is extremely scarce in comparison to other topics, such as Concurrent Engineering (CE) and Total Quality Control/Management (TQC/M). Only one volume of monograph was found to cover this topic systematically (Monahan, 1995). Other studies include standards such as American MIL-STD-973. A few design and manufacture texts also included the ECM topic as a chapter (Leech and Turner, 1985). The review by Wright (1997) is, by far, probably the most comprehensive literature review, since it is based on fifteen relevant papers.

In total, about three dozen articles have been retrieved on ECM. Some articles discuss the topic in reasonable depth. These articles can be divided into four categories: 1) survey and review articles (Maull et al, 1992; Boznak, 1993), 2) reports on industrial case studies (Saeed *et al.*, 1993, Hegde *et al.*, 1992; Balcerak and Dale, 1992; Harhalakis, 1986; Watts, 1984), 3) a few articles providing implementation frameworks (Dale, 1982; Harhalakis, 1986) and 4) strategic conceptual guidelines (Diprima, 1982; Nichols, 1990; Reidelback, 1991).

In an international survey conducted among American and European companies in 1988 (e.g., aerospace, defence, textiles, electronics, consumer products, construction, utility, ship repair, and foundry), Boznak (1993) found that the cost of change experienced in these companies underscored their difficulty in reducing product costs. The number of changes, as recorded from the respective companies, varied from 2 to 1,000 monthly changes, with an average of 330 design changes per month. Administrative processing costs, from small firms to FORTUNE 500 companies, average US$1,400 per change. This corresponds to an annual administrative processing cost ranging from US$3.4 million to US$7.7 million.

Numerous industrial case studies were found on the ECM subject. Saeed *et al* (1993) have conducted an intensive case study within an American FORTUNE 500 manufacturing company. Hegde *et al.* (1992) have carried out a field investigation on the impact of engineering change orders (ECOs) on completion time of jobs in repetitive manufacturing environments. Balcerak and Dale (1992) have examined ECM in an automotive manufacturer and made a number of specific suggestions on EC classification, prioritisation, and effectiveness evaluation. Harhalakis (1986) carried out industrial case studies on engineering change management for made-to-order products. Watts (1984) expounded his experience in reengineering the EC control process in electronic product manufacture. Dale's early work in 1982 was perhaps one of the best papers in providing a comprehensive framework with specific and good practice guidelines (Dale, 1982). Harhalakis (1986) derived from the case studies on his company and outlined an ECM system. These contributions not only described ECM activities, but also identified responsibilities.

Despite its scarcity, the message brought up in the literature is clear: EC is a serious issue demanding considerable attention and ECM is of major concern to most companies that design and manufacture products. This has been confirmed in a survey on UK manufacturing industries conducted by the authors (Huang and Mak, 1997b). However, the limited references also show that ECM is generally under-researched and ECM deserves much more attention, in view of its importance to industry.

Most of the literature has been based on paper-based ECM systems. Although some companies seem to have well-structured comprehensive ECM documents, paper-based systems are often incapable of managing ECs with sufficient effectiveness and efficiency. For example, the systematic procedures clearly defined in a document may not be applied in practice. Some companies adopt a system where paper-based EC packages are circulated according to certain flowchart, and the EC package is passed onto the next function after the preceding function finishes processing it. This leads to another limitation, i.e., mono access for one user at a time, and is often evident in a Sequential Engineering environment. Such sequential processing results in excessive throughput time. The problem worsens if the EC package must be returned to the previous functions for re-processing. One alternative is to produce multiple copies of the EC package for simultaneous distribution to all the parties concerned. However, this approach not only consumes more paper and involves more paperwork, but also creates difficulties in collating results.

In view of the aforementioned limitations of the paper-based ECM systems, Information Technology (IT) has been introduced to overcome these limitations. These computer aided ECM systems are generally classified into 3 types (Huang and Mak, 1997c): 1) primitive applications, such as the use of word processors to prepare EC documents and spreadsheets to record EC data; 2) applications specially developed to support basic EC activities, including requesting and recording ECs; 3) the recent Product Data Management (PDM) and Enterprise Resource Planning (ERP) systems. They provide more comprehensive functionality than just managing ECs. With the aid of ECM software packages, the implementation of ECM can be more convenient, since the parties involved can communicate easily and consistently, and information can be accessed, amended and updated at any time through a computerised environment.

It is found (Huang and Mak, 1999a) that a very small number of manufacturing companies have introduced computer aided ECM systems, although many have expressed interest in adoption. Some large companies have invested heavily on PDM technology, but the systems seem to be too complex even for them to initiate the ECM functionality. Another group of large companies has developed their own ECM systems to record and process EC data. However, these software packages are standalone and accessible only to a single user at a time. Simultaneous access by multiple disciplines, especially geographically disparate ones, is usually infeasible.

These observations have induced the present study to explore the web technology to support EC activities on the Internet/intranets. Indeed, the web technology has recently been adopted by numerous major research projects. An overall review of the applications is presented in Huang and Mak (1997c) Among the most notable is the MADE (Manufacturing Automation and Design Engineering) program initiated in 1992. The MADE program has been extended now to the RaDEO (Rapid Design Exploration and Optimisation) program (http://radeo.nist.gov/radeo/) which employs the Internet technology extensively for tele-design and manufacture. Recent initiatives on Agile and Virtual Manufacturing (http://agile.cimds.ri.cmu.edu/ and http://www.isr.umd.edu/Labs/CIM/vm/) also make extensive reference to the Internet/Intranet technology in general and the web technology in particular.

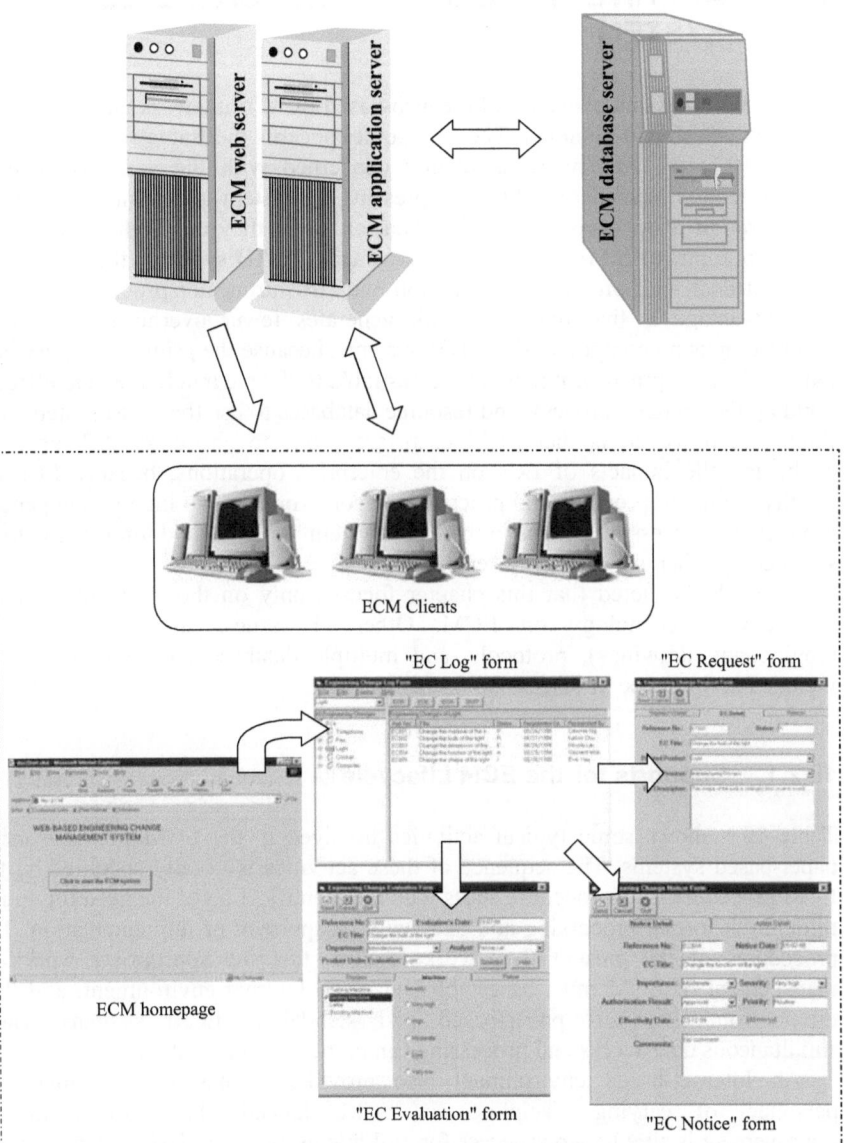

Figure 10.1 Overall architecture of a web-based ECM system.

10.2.   ARCHITECTURE OF THE WEB-BASED ECM SYSTEM

Figure 10.1 shows an overview of the web-based ECM framework, and 2 points of clarifications should be noted. The first point concerns the framework's scope. In Figure 10.1, the framework is mainly concerned with the various functions directly related to managing ECs. At present, the system is not intended for the incorporation of extensive enterprise data, although they are highly relevant to ECM. In contrast to Product Data Management (PDM) systems, the framework does not deal with product configuration and structuring nor process workflow. One advantage is that the framework generates fewer overheads than those involved in using comprehensive PDM systems, because the primary purpose is to manage ECs. In practice, it may not be desirable to devote much time and effort to build up the product, process, and resource databases to use the ECM system. The limitation, however, is that ECM users are prone to encounter difficulties in evaluating the impacts of ECs on the enterprise operations, because ECs are usually related to products and processes. A compromise is to incorporate general product and process data items in the ECM system, with detailed information to be supplied by other specialised systems.

It should be noted that this chapter focuses only on the application of the Internet/web technology on ECM. Other IT issues, such as bandwidth (engineering drawings), protocols, and multiple database connectivity are not included in the scope of discussion of this chapter.

10.2.1.   Supports for the ECM Lifecycle

Table 10.1 shows some typical activities involved in the ECM lifecycle using paper-based systems. The sequence of these activities is usually specified by the ECM procedure. The procedure starts with the identification of the need for an EC through its formal processing and ends with disapproval or implementation. The procedures of the present framework requires that the appropriate workflow management of ECM must be established in an Internet environment, and thus differ from those of paper-based and standalone ECM systems where simultaneous data access and processing cannot be accommodated.

An Internet-based environment also provides a new way of involving personnel of varying disciplines in ECM. Indeed, the most significant improvement is simultaneous access for multiple users, regardless of geographical disparity. For example, a message can be broadcast to all the concerned parties upon a formal proposal of an EC, where the concerned parties can instantly receive and start evaluating the proposed EC from their perspectives. The evaluation results are retained in the ECM database. If there is an EC co-ordinator, he/she then collates all the results and prepares appropriate documentation for the ECM Board to approve or to disapprove the requested EC. Such board meetings can take place virtually, and participants can express their opinions and cast their votes electronically. This type of functionality has now become available in some group decision support systems (Romano *et al*, 1998).

Table 10.1 Typical ECM lifecycle activities

- Receive EC application
- Raise EC proposal
- Forward package to co-ordinator
- Receive and record EC package
- Request stock check and enter data on EC note and information sheet
- Complete stock check and return to co-ordinator
- File a copy of stock check form
- Forward package to Purchasing
- Record data on EC note and information sheet, return package to co-ordinator
- Forward to computer file via Product Engineering
- Forward package to Industrial Engineering
- Complete package and return to co-ordinator
- Forward item's master change to computer file via Product Engineering
- Inform departments that require information
- Forward package to Cost Office
- Record data on EC note and information sheet and return to co-ordinator
- File package
- Submit package to Management Committee
- Appraisal of proposed ECN
- Inform all departments of ECN status and forward package to Product Engineering
- Activate ECN
- Forward package to co-ordinator
- Request stock check, complete requisitions
- Complete stock check, return to co-ordinator
- Forward requisition and package to Purchasing
- Submit purchase orders
- Confirm and return package to co-ordinator
- Inform all departments that the ECN is activated
- Inform Quality Control and co-ordinator upon availability of sample approvals
- Inspect sample approval
- Inform co-ordinator of findings
- Forward illustration to Marketing
- Arrange for printing

After Dale (1982)

As far as documentation is concerned, it is very much simplified in comparison with paper-based and standalone ECM systems. Firstly, the multimedia capability of the web enables the development of user-friendly and sophisticated documentary interfaces, and the system users should not encounter any significant changes at the user interface level. Secondly, the circulation of hard copies diminishes because access becomes available any time from anywhere through a general-purpose web browser. Hard copies can always be retrieved by individual users, if they prefer such formats.

10.2.2.   Components of the Web-Based ECM System

The web-based ECM system follows the Microsoft's Windows DNA (Distributed interNet Applications) three-tiered model of web applications. Four components are included in the web-based ECM system, namely, an ECM database server, an ECM web server, an ECM application server, and a number of ECM clients. These components can be geographically dispersed anywhere in the world (or within the corporation), but ECM is applicable to them all as long as they are all connected to the Internet (or the corporate Intranet). The next section discusses in detail the implementation and the deployment of these components, whereas the rest of this section gives a brief synopsis of each of these elements.

ECM database server

The ECM database server is the very first data source tier in the three-tiered model of web applications, and can be a database server installed on a common or different machine from that of the ECM web/application server. A relational database management system with Structured Query Language (SQL) is selected for this project. EC data are represented by numerous data tables, and each table is defined by a number of fields. EC data entries are maintained as records or rows in the corresponding data tables. The lifecycle of ECs encompasses the following stages: raising, processing, approving, notifying, implementing, and auditing. Accordingly, the ECM database records the data required and produced at these stages. A model of the ECM database for the ECM lifecycle is presented in Section 10.4.1.

ECM web server

The ECM web server, the middle tier, is an ECM web page at which the functional components are deployed as forms. Each form contributes certain functionality that supports some or more ECM activities along the ECM lifecycle. An ECM homepage or the start-up form includes links to a number of ECM functional forms, including Engineering Change Log (ECL), Engineering Change Request (ECR), Engineering Change Evaluation (ECE), and Engineering Change Notice forms. Users can make use of these forms to access or retrieve EC data and to conduct a variety of analyses.

ECM application server

Together with the ECM web server, the ECM application server is another component of the middle tier in ECM web applications. In practice, both the ECM web and application servers can be installed together with great physical proximity on the same machine, or separately on two different machines. The ECM application server is tasked with two main functions in ECM. The first function is to collect and process all the data delivered from the client already downloaded from the ECM web server. The second function is to communicate with the ECM database server in order to retrieve the data from or to make changes in the ECM database.

ECM clients

The client tier includes a set of ECM clients because the ECM clients cannot be actually classified as part of the web-based ECM system, and they have no practical relation with the ECM at all prior to connection. They form one of the constituents of the system only when they are connected to the ECM web server.

In theory, an infinite number of ECM clients can exist and they can use the services provided by the ECM web server. For example, customers and suppliers can make EC requests for product changes outside the company at different locations through the web browsers. Internal departments can also use the web browsers to make EC requests at any time and location. The data collected from the clients will be submitted to the ECM web server, and the ECM database server will be invoked to manipulate the data in the assigned database. These users can access the relevant ECM web pages to request a specific EC, to inquire about the status of a previously requested EC, to evaluate an EC, and to receive EC implementation notices.

Once the client browser is connected to the ECM web server, the ECM system is downloaded to the client machine. Four functional ECM forms, which are incorporated with the downloaded ECM system, are delivered to the client machine as well. A brief description of these forms is given as follows:

- *EC Log form.* It presents an overview of the ECs that have already entered into the ECM system, including the major information of ECs, such as EC reference number, EC title, the date of EC request, and the name of the originator. This information is useful to the users to understand easily all ECs or a selected EC. Above all, the EC Log form provides a medium for users to retrieve other ECM forms.
- *EC Request form.* It offers the users a means to propose an EC, and to retrieve the information of a selected EC request from the ECM database and to view it on the EC Request form.
- *EC Evaluation form.* It analyses a requested EC in terms of its impacts on products, processes and other organisational and/or operational aspects. The results delivered from the EC impact analysis of a chosen EC can also be retrieved from the ECM database and shown on the EC Evaluation form

- *EC Notice form.* It notifies the disciplines about the approval of ECs. The Notice form also allows the users to acquire and to view the information relevant to a specific EC Notice form.

10.3.   SYSTEM DEPLOYMENT AND IMPLEMENTATION

This section discusses further on the construction of the ECM database, the deployment of the components of the web-based ECM system, and the implementation of the web-based ECM system.

10.3.1.   ECM Database Model

In an experimental implementation, it is found that the ECM database server resides in a geographically disparate machine from the client machine. It retains the data relevant to ECs throughout the entire ECM lifecycle. Several data tables, classified under two categories, are created to represent the EC data in the ECM database, namely, the basic ECM data tables and supplementary data tables.

The basic ECM data tables stores data that are closely relevant to the requested ECs. Four basic ECM data tables are created in this project. One of the tables records the main EC data along the ECM lifecycle, while the rest store EC data in the stages of request, evaluation, and notice of the ECM lifecycle. The four main data tables are briefly described below:

- *The ECs table* is the master table that records all ECs. The main items include a unique EC reference number, its title, its description, its status, and so on. Data records in other tables are all linked or attached to entries in this table.
- *The Reasons table* offers descriptions of the reasons of requiring particular ECs. The reasons are rated according to their importance. More than one reason can be used to justify one EC, and these reasons serve as criteria to prioritize the ECs together with their impacts.
- *The Impacts tablet* provides the evaluation of the impacts of particular ECs on products, processes and resources in terms of operation and organisation. One EC may affect more than one item from different perspectives. Impacts can be defined in either qualitative or quantitative terms or both. Indeed, impacts are assessed alongside with reasons to classify and prioritise the EC requests.
- *The Actions table* records actions that are required for particular ECs. Actions are described in terms of action explanations, the starting and finish times, the responsible personnel, etc.

In addition to the above four primary data tables, a number of supplementary data tables are also adopted in the system. These tables do not contain any information about ECs but, instead, list useful information necessary for EC impact analysis and implementation planning. For example, The Organisation table keeps records of personnel and their responsibilities in ECM. In addition, product, process and resource information is also required to indicate the victims or beneficiaries of ECs. For example, the impacts of an EC on a manufacturing process or operation can be evaluated according to tooling, equipment capability, process parameters, and time, etc. All these data must be stored in the database when they become involved.

10.3.2.　System Deployment

There are three basic ways of deploying a web-based ECM system on the Internet: on the server, on the client, and distributed on both sides. The first two ways are shown in Figure 10.2. If the ECM system is deployed on the client machine, the client downloads the ECM system as soon as the client browser is connected to the server page, as shown in lines 1 and 2 in Figure 10.2(a). Computation takes place in the client machine, and there is no further need to communicate with the web server. The remote ECM database server must be established by all the clients, as indicated by lines 3 and 4 in the figure. This means that the user has the freedom or is required (MEANING CONTRADICTORY) to specify a remote data source. The most significant advantage is that the user can interact with the ECM system as frequently as necessary without incurring any communication overheads with the server. A disadvantage, however, is that the client downloading time may be long because the clients are "fat".

In Figure 10.2(b), the server-side deployment requires the client browser to collect only input data from and to display outputs to the user, as indicated by lines 1 and 4 in the left-hand-side of the figure, while computation is carried out by the server machine. The ECM application server will be entirely responsible for communicating with the ECM database server, as indicated by lines 2 and 3 in the figure. This method is not suitable for applications where the user has to interact with the system frequently, because the communication overheads between the server and the client may become expensive. However, the downloading time of the clients is short because the clients are "thin".

Taking the pros and cons of these deployment methods into consideration, it was decided that the "fat" client approach be adopted initially to experiment with the ECM prototype. Another reason for this choice is that the testing of the underlying ECM methodologies becomes easier during its incorporation into its corresponding components. The "thin" client implementation was proposed to begin after sufficient insights on the "fat" prototype have been gained in the near future.

10.3.3.　System Implementation

In recent years, numerous technologies are available in providing a set of features and tools that can be used to implement the application systems in the Internet environment, for example, Java, ActiveX, and so on. In this project, ActiveX technology is selected. The reason is that the ActiveX technology is capable of improving the interactivity between the user and the system, thus achieving the objectives established for this project.

A number of components are provided by the ActiveX technology. For example, ActiveX controls are objects embedded in the web page. ActiveX documents can be viewed on the client machine. ActiveX code components can provide the functions of adding, deleting, and updating the data.

At the initial stage of this project, the system components were developed in the ActiveX codes. The main ECM functionality was developed and could be implemented on the same machine. Then, the system was further developed into the client and server architecture for deployment on the client-side. The ECM functionality can only be implemented on specific client and server machines. Thereafter, the approach of deployment was switched to server-side. The client and server machines can communicate with each other through the Internet environment in order to implement the built-in ECM functionality. Despite the long-distance separation, the system experiences no discrepancy. The system is incorporated into the ActiveX document so that the client can download the system through the Internet.

10.4.   WEB-BASED ECM FACILITIES

The web-based ECM system supports three main stages in a typical ECM lifecycle. They are (1) when an EC is identified and formally requested and awaiting formal approval, (2) during the evaluation, classification, and prioritisation of an EC, and (3) after an EC is approved when it is released for implementation. In practice, the web-based ECM system provides four forms that aid the management of ECs in these three stages. They are the EC Log (ECL), EC Request (ECR), EC Evaluation (ECE), and EC Notice (ECN) forms. Potential users only need to connect to the ECM web server to retrieve the relevant form to make their contributions. The content of each ECM form and the prototype implementation of these pages are discussed in the remainder of this section.

10.4.1.   Web-Based Log of Engineering Changes

Engineering Change Log (ECL) form is a start-up form of the ECM system. Its use is to present an overview of the ECs which have already entered into the system to serve as a favourable device to assess other ECM facilitates. Figure 10.3 shows a sample screen of an ECL web page.

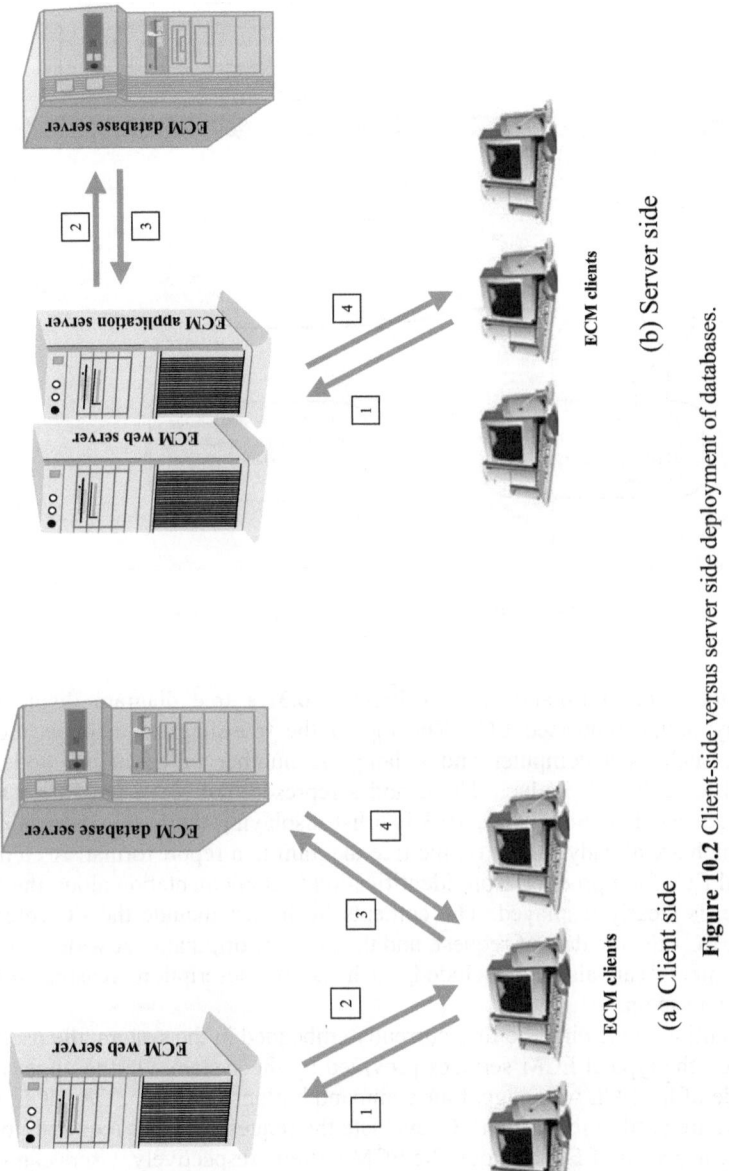

Figure 10.2 Client-side versus server side deployment of databases.

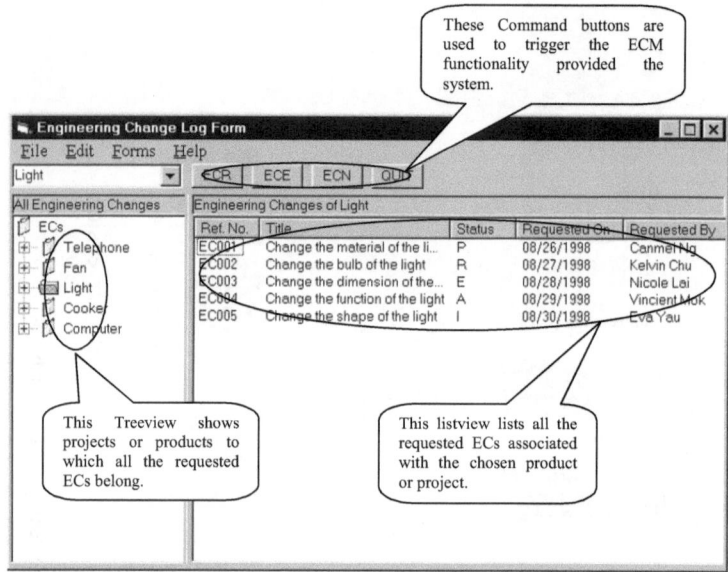

Figure 10.3 Sample screen of an ECL form.

On the lower left-hand side of Figure 10.3, a tree diagram displays the hierarchy of the requested ECs. The top of the tree diagram displays the end product, such as a computer and a lamp. A number of cascading nodes are attached to each end product. These nodes represent the requested ECs. On the lower right-hand side of Figure 10.3 is a list displaying the basic contents of the ECs, which are already shown on the tree diagram in a report format. A complete list of all ECs for a product, from identification to implementation along the ECM lifecycle, is clearly displayed. The contents in the list include the EC reference number, EC title, the date of request, and the name of originator. A wide variety of other contents can also be included, such as EC description, reasons for EC request, and so on.

In addition to viewing the ECs currently embedded in the system, the users can also enjoy the typical ECM services provided by the system. On the upper right-hand-side of the ECL web page, four command buttons, namely, ECR, ECE, ECN, and Quit, are used to request an EC, evaluate the requested EC as required, inform about the approval of EC, and quit the ECM system, respectively. Users can select one of the functions according to their own needs by clicking the corresponding command button.

10.4.2. Web-Based Request of Engineering Changes

Engineering change request (ECR) form is a data acquisition form for a formal proposal of an EC that has been identified by someone (originator or proposer) through some sort of analysis. Once an EC enters the web-based ECM system, it must be dealt with and cannot be ignored. Figure 10.4 shows a sample screen of an ECL web page.

The ECL form contains three tab pages, namely, "Originator's Detail", "EC Detail", and "Reason", and each tab page is responsible for a particular aspect in the application of the EC request. The "Originator's Detail" tab page contains a number of textboxes to collect the originator's information, such as the identity of the originator for the EC and the discipline from which the originator comes from. In the "EC Detail" tab page, the textboxes are also included to acquire the specific information about the proposed EC, like what the EC is about and the product(s) and process(es) relating to the EC. The "Reason" tab page consists of a list of reasons for proposing the EC request and also the significance of each reason.

The data collected in the ECR form are recorded in the ECM database server for future use. Several events may be triggered following submission of the ECR page. For example, a message can be automatically broadcast to all members on the ECM Board. Alternatively, the message may be forwarded by the EC co-ordinator, if any, to draw the attention of members in the ECM Board and set a deadline for evaluating the EC. All the Board members can retrieve the EC data and take necessary actions as required.

10.4.3. Web-Based Evaluation of Engineering Changes

An engineering change evaluation (ECE) form is a data acquisition form for collecting the data after conducting the impact analysis on a requested EC by various disciplines. Only those data that are essential to the EC impact analysis are acquired to facilitate data collection and processing of ECM.

There are two approaches towards the EC impact analysis: quantitative evaluation and qualitative evaluation (Yee *et al*, 1998), and these 2 approaches differ in terms of ease of use and accuracy of delivering the results. The quantitative evaluation is the preferred approach because it provides the exact values of the EC impact. However, it is very difficult to determine the values of the EC impact because of the lack of information. The qualitative evaluation is easy to use, yet unable to give an accurate estimation on the EC impact. In practice, these approaches are often combined together, thus semi-quantitative evaluation is usually used to evaluate the EC impact on the proposed web-based ECM system, and the items affected by the requested EC are checked quantitatively first. Since then, each affected item is assessed qualitatively with a pre-defined severity reference to determine the severity of its impact to the business process.

In general, an EC poses effects on a series of activities in various business processes, such as manufacturing, purchasing, design and development, and so on.

In this connection, a requested EC must be assessed from different perspectives to ensure clarity. In other words, different business processes may require different kinds of EC impact analysis. For example, the operational process, machine, and fixture are considered the potential impacts on the manufacturing process, whereas the drawing and parts are taken to be the potential impacts on the design and development process. A more detailed example will be given in the following for the illustration.

Figure 10.5 shows a sample screen of an ECE form. This form is particularly designed to perform semi-quantitative EC impact analysis in the manufacturing process. It is basically made up of two parts: general evaluation information and specific EC impact analysis. At the top of Figure 10.5, a number of textboxes and combo boxes that are included to collect the basic information for the EC impact analysis, for example, the name of analyst and the date of performing the EC impact analysis.

The lower portion of Figure 10.5 is designed specially for carrying out the EC impact analysis in the manufacturing process. As mentioned before, the EC in the manufacturing process may induce impacts on the associated manufacturing processes, tooling, and fixtures. The first step of the semi-quantitative EC impact analysis is, therefore, to evaluate the requested EC qualitatively in these aspects. The evaluation is done by checking a list box on the lower left of the figure which lists the potentially affected items of a requested EC. Once an item that is affected by the subjected EC is identified, the quantitative evaluation follows. The degree of severity of each affected item is quantified objectively by a pre-defined reference. An optional frame, called severity, is popped up on the right-hand side of the tab page after an item is clicked in the list box. The severity frame provides users with several ranks to assess the severity of the EC. Similarly, semi-quantitative evaluations on other business processes can be performed.

The data collected in the ECE form are recorded in the ECM database server for further processing in the later stages of the ECM lifecycle. After the ECE form is submitted, a list of activities will follow. For instance, the EC co-ordinator will centralise all the information of the submitted ECE forms from various disciplines across a company. Then, trade-off analysis can be carried out, where applicable, among concerned disciplines to optimise the outcomes delivered by the EC. Members in ECM Board may meet together and retrieve the recorded data to decide whether the requested EC is worth and necessary.

10.4.4.   Web-Based Notice of Engineering Changes

An engineering change notice (ECN) web page is a data transmission form to disseminate notices on approval of EC and actions that follow to the various disciplines across a company for various purposes. The purposes are to gain support from the entire company, and more importantly, to prepare the concerned disciplines to implement the approved EC.

A sample screen of an ECN web page is shown in Figure 10.6. It contains two tab pages, namely, "Notice Detail" and "Action Detail". The "Notice Detail" tab page contains a number of textboxes used for acquiring details of the authorisation, for instance, the importance, severity, and priority of the EC. In the "Action Detail" tab page, the textboxes collect the information relevant to the action schedule, and a list is used to show the identified actions in a report format.

The data obtained from the ECN form are retained in the ECM database for future use. The concerned parties prepare their own disciplines to implement the approved EC. For example, the manufacturing discipline may change the tooling; the design discipline may release an amended design for the product. The EC co-ordinator may check whether the concerned disciplines take their individual actions according to the action plan.

10.4.5. Typical Procedure of Using the ECM Facilities

The procedure for utilisation of the web-based ECM system is dependent on the way of deploying ActiveX code components. In the previous sections, it was noted that ECL, ECR, ECE, and ECN codes are deployed on the server-side as code components. Potential users should therefore connect their machines to the ECM web server and retrieve the relevant web pages that activate these code components. Upon execution, these code components act in the same way as standalone systems on the client machine.

The process of using this web-based ECM system is as follows. First, the user must use a web browser (Internet Explorer) to connect to the ECM web server by typing the URL address. The ECM web server then sends the ECM homepage to the client's web browser and, simultaneously, the ECM system components attached to this web page are also downloaded automatically. Just a click of a button on the web page will activate the ECM system. The ECM system is already operational, behaving just like a standalone system.

The first form appearing is the ECL form. A list of ECs is displayed in the main window. Once an EC is chosen, command buttons are enabled. The appropriate button is clicked. The ECR button is used if the user wants to make an EC request or modify an existing EC request. After all the details are entered, the form is submitted to the ECM application server that will, in turn, save all the details in the ECM database. ECE and ECN buttons are used to activate the ECE and ECN facilities to evaluate the EC impacts and to issue EC notices, respectively.

One point worth mentioning is that the web-based ECM system provides two operational modes. One is the editing mode and the other is the view-only mode. In the editing mode, the user can use the ECM facilities to change the contents of the EC details. In the view-only mode, the user can retrieve the EC data from the ECM database of a specific EC on a particular form, view the retrieved data on the corresponding form, but has no authority to change the data. This should be controlled at the beginning when the user logs onto the system. More developments on utility facilities as such can be built in to enhance the ECM system.

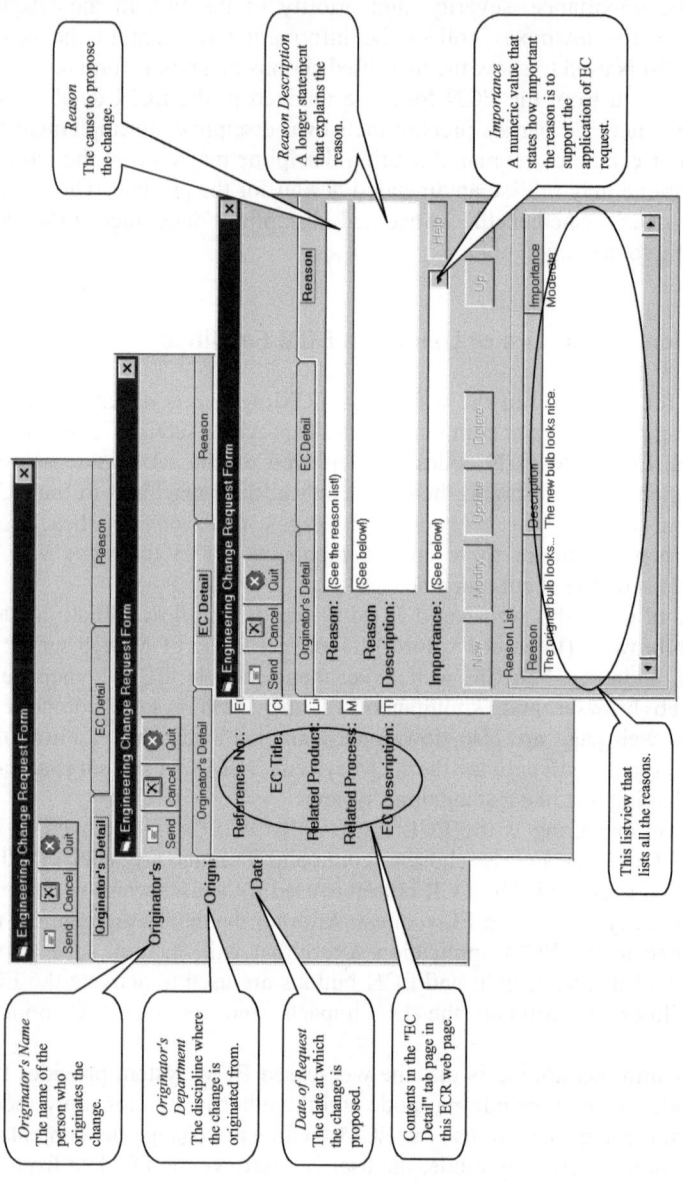

Figure 10.4 Sample screen of an ECR form.

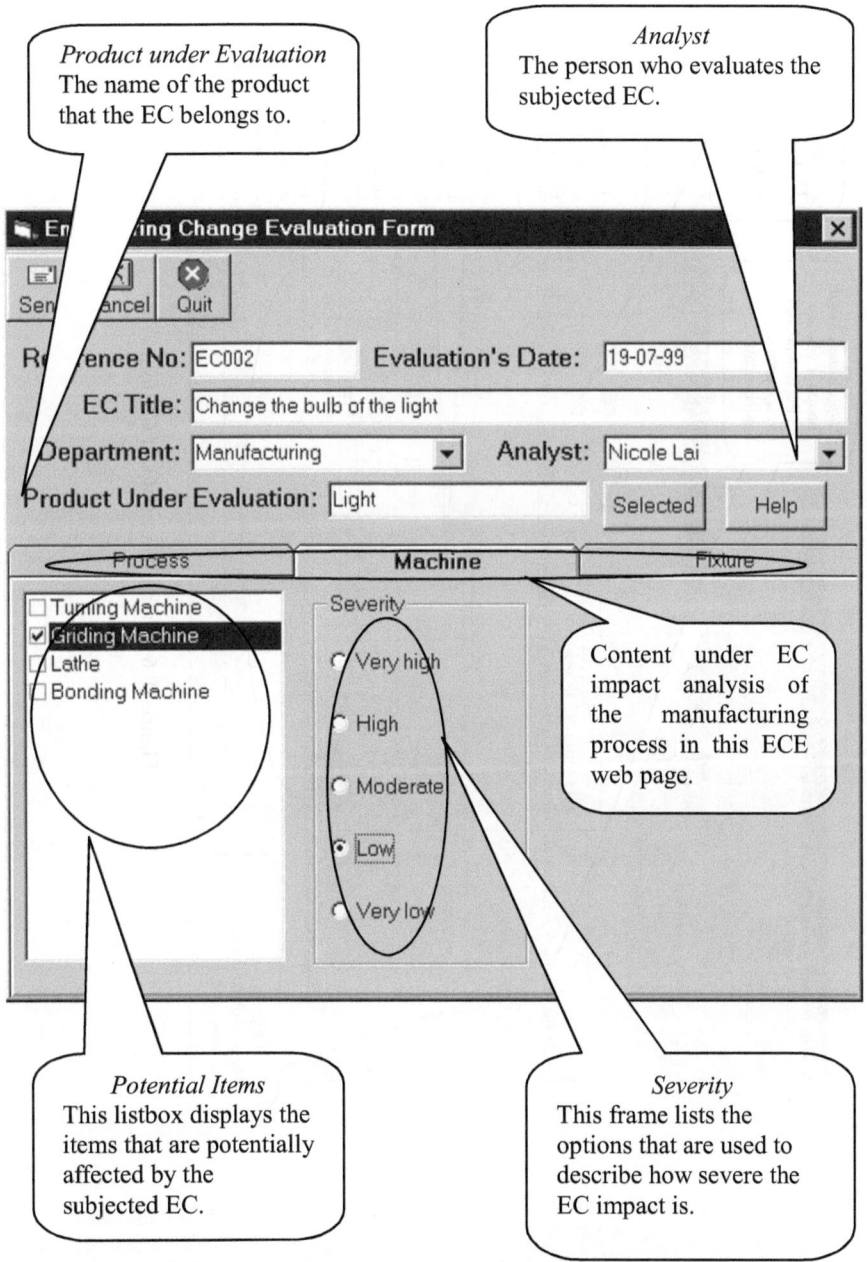

Figure 10.5 Sample screen of an ECE form.

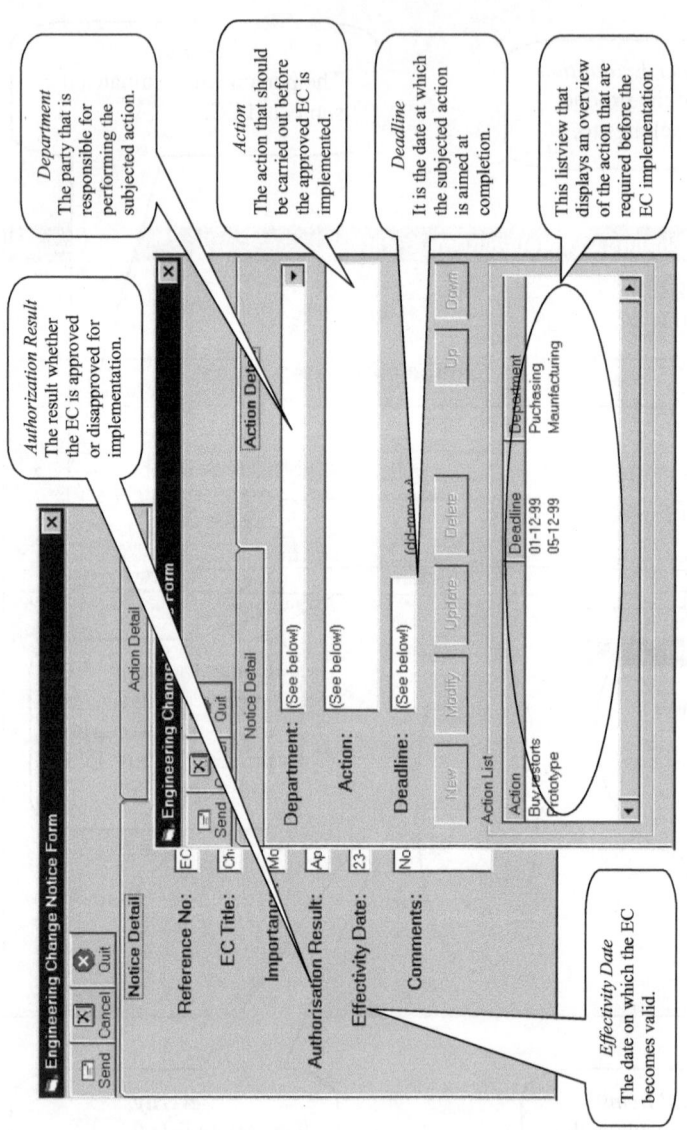

Figure 10.6 Sample screen of an ECN form.

10.5. SUMMARY

This chapter has discussed on the application and development of web technology on computer systems for engineering change management (ECM). Information regarding organisation, procedure, and forms are easily built into the system database. Facilities are provided to retrieve the data and display them in various formats suitable for particular applications in managing ECs.

A web-based ECM system offers a number of advantages over paper-based and standalone computer ECM systems. First, the amount of paperwork is reduced to a minimum level. Second, the throughput time is significantly reduced, mainly because of the third advantage of simultaneous data access and processing which a web-based ECM system can accommodate, unlike the single user access for paper-based and standalone systems. Finally, EC data are shared and communicated among all parties immediately after entry into the system.

The web-based ECM system mainly supports basic ECM functions and activities at present. Further developments are possible both within and across the defined scope. Within the defined scope, it is intended that more ECM decision supports be incorporated, as the present research progresses into the development of methodologies for EC impact analysis and effectiveness planning, EC evaluation, and EC classification and prioritisation, etc.

The scope of the ECM system can be extended to incorporate the facilities of comprehensive PDM systems. The interfaces with other decision support systems, such as CAD, CAPP, MRP, etc., need to be addressed in the long run. A short-term compromise is the inclusion of outline information, but not detailed information, of products, processes, and resources should be integrated into the system. Such outline catalogues are expected to provide pointers to relevant data sources or application systems.

ECM involves the handling of intensive data of products and processes. In practice, it is difficult to collect the appropriate data and analyses can be time-consuming. To overcome these problems, ECM web pages are included as a part of the ECM system. The contents in the ECM web pages of a web-based ECM system determine the scope and the effectiveness of its functionality. The formats in the ECM web pages affect the speed and efficiency of its use. Well-structured ECM web pages are easy to use.

In addition to the functionality, the prototype system highlights a few key issues in terms of system implementation and deployment. As expected, the downloading time of the so-called "fat" ECM web pages are relatively long, especially for the first time. In refining the next version, experiments will be done on the "thin client" approach. Another issue worth highlighting is that it is not straightforward to customise the system to suit the specific needs of different manufacturing organisations. The industrial and literature surveys of the present study have indicated that different organisations seem to operate ECM systems with slight variations, and it is thus necessary for a computer system to adapt itself to suit this evolving situation.

11

WEB-BASED SUPPORT FOR
EARLY SUPPLIER INVOLVEMENT IN
NEW PRODUCT DEVELOPMENT

In the last decade, the significance of concurrent engineering and supply chain management has been widely acclaimed, as evidenced in the bountiful support to research and project development from both the government research councils and industrial bodies all over the world. Indeed, one common finding is that greater benefits can be accrued from supplier involvement in the more initial stages of new product development process. The rationale is that suppliers usually possess vital product and process technology that can lead to improvements in product design and the New Product Development (NPD) process itself. For example, it has been found in a cross-national study of 29 NPD projects that much of the Japanese advantage in concept-to-market time was attributable to supplier involvement in the NPD process (Clark, 1989). Indeed, it is not exaggerating to say that there exists infinite potential from the suppliers in impacting on the quality and cost of new products (Burt, 1989).

Early Supplier Involvement (ESI) has emerged as a good practice in NPD to ensure that the positive impacts are maximized while minimizing the negative impacts, on the general understanding that ESI is beneficial to both the buyers and suppliers. Typical benefits (Bonaccorsi and Lipparini, 1994) included reduced development costs, early availability of prototypes, standardisation of components, visibility of the cost-performance tradeoff, consistency between design and supplier's process capabilities, decreased engineering changes, higher quality with fewer defects, consistency between product tolerances and process capabilities. Some other specific logistical advantages include refinement of the supplier's processes, availability of detailed process data, reduced time to market, instant identification of technical problems, shortened supplier's process engineering time, acquisition of suppliers' production capacity, and improved supplier innovation.

ESI in NPD is not only imperative but also a big challenge (Dowlatshahi, 1997; Twigg, 1998). For example, Dowlatshahi (1997) identified a number of propositions for ESI from a theoretical point of view. Fleischer and Liker (1998) identified supplier roles in product development and the mechanisms of ESI through interview and mail surveys, in order to study the application of ESI in industries. Although the significance of these suggestions are indisputable, empirical investigations within world-wide manufacturing industries indicate that there are wide variations in ESI adoption, and that different factors have varying effects on the choice of the approach of adoption (Bidault et al, 1998).

Furthermore, a recent survey conducted by Boston et al (1998) with UK manufacturing industries indicates that only 36% of the organisations had formal procedures in place for the management of supplier information. Over 60% of those companies maintained supplier literature within a global library, and their supplier literature was merely classified according to its original format (i.e., catalogues, handbooks, data sheets, etc.) or the supplier names. Thus, an engineering team that is familiar with what the supplier produces is required in order to transform this classification system to be of value.

The main purpose of this chapter is to address the gap between the importance and the difficulty of adopting the ESI in NPD approach. A web-based framework, called WeBid, is under development to promote and to facilitate the ESI in NPD approach in practice. This chapter summarises the results of the initial investigation. Section 11.1 presents the needs for early supplier involvement in new product development. Section 11.2 presents an overview of the WeBid methodology and framework. Sections 11.3, 11.4, 11.5 and 11.6 focus on the Supply Explorer, the Bid Explorer, the Partnership Explorer, and Share Explorer respectively. Section 11.7 briefly summarises implementation issues.

11.1.   EARLY SUPPLIER INVOLVEMENT (ESI) IN NPD

This section identifies new research challenges in ESI in NPD based on a brief review of the literature and practices.

11.1.1.   Early Supplier Involvement

Twigg (1998) compiled a typology of supplier involvement in product development from various sources. This typology provides a basis for separate consideration of the requirements and participation of suppliers with different inputs. Figure 11.1 presents a different typology that indicates supplier involvement at different stages of the product development process. Only four main stages are represented in the diagram. Indeed, it is vital though extremely difficult to involve suppliers early at the product specification stage (Karlsson et al, 1998). At the concept design stage, suppliers help identify the latest technologies to be incorporated into a new product. Suppliers contribute to the detailed design by providing solutions to the component and part designs and by selecting the most suitable materials and catalogue components. Suppliers can also assist in the "Make or Buy" decision-making when production design begins. Infrastructure suppliers provide the most capable tooling, fixturing and equipment. Above all, suppliers may be involved throughout the entire NPD process to carry out Design for Manufacturability analyses to ensure that the product is delivered effectively and efficiently.

The word "early" in Early Supplier Involvement can be interpreted from different perspectives. Supplier involvement in product design specification (PDS) is considered early involvement in the sense that PDS is an early stage in the product development process. Another interpretation is that a supplier is involved earlier than must be actually involved originally. For example, an equipment supplier is to be involved early at the stage of production design or even earlier, at the stage of detailed part design, if a manufacturing facility must be purchased to make certain parts at the production stage. Such early involvement helps the development team to establish suitable tolerances according to the equipment capabilities to be acquired.

11.1.2. New Challenges

Considerable attention has been given to ESI in NPD, as both the Concurrent Engineering (CE) and Supply Chain Management (SCM) research communities have placed emphasis on this issue. However, numerous unique issues have yet to be addressed further. Firstly, there is a need for a new model of supply chain to support ESI in NPD. Previous supply chain analyses and modelling have focused on the entire chain (Beamon, 1998) or individual member companies along different levels of the chain (Choi et al, 1996). In order to capture the full complexity of the process of supply in a more holistic and strategic view, Lamming (1999) presents the use of the term 'supply network' to define the process of supply which involves complex non-linear links among inter-connecting supply entities. These have been reflected in some well-known supply chain models, such as SCOR proposed by AMR Supply Chain Council (www.supply-chain.org), and CPFR by Voluntary Inter-Industry Commerce Standards Organisation (www.cpfr.org). However, these models are limited because they do not analyse and model the interfaces between the customers and suppliers. In particular, they do not reflect the customers' nor the suppliers' product development processes.

Secondly, a new method for supplier selection is needed to support ESI in NPD. The importance of supplier selection stems from the fact that it commits resources while simultaneously impacting not only such activities as inventory management, production planning and control, cash flow requirements, and product quality, but also product design and development (Helper, 1995). The literature in vendor or supplier selection continues to grow rapidly in areas such as vendor or supplier attributes and performance metrics (Verma, 1998; Choi, 1996), and decision models (Holt, 1998; Ghodsypour, 1998; Vokurka, 1996). Most previous research in the area of supplier selection and supplier evaluation emphasizes on conceptual and empirical decision support models designed mainly for purchasing managers. They may suffer from one or more shortcomings, such as being mathematically too complex, too subjective, excessive supporting data, etc (Holt, 1998). Practitioners are, nevertheless, looking for a methodology with simple applications and instructions yet producing reasonably accurate results. However, the literature reveals that supplier attributes or performance metrics for conventional supplier selection do not seem to include design specifications that are usually provided in the early stages of product development.

Thirdly, supply development is normally based on competitive tender/bidding, despite the fact that partnership development is encouraged to increase efficiency in purchase. Competitive tender/bidding is considered adversary, and undermining collaborative partnerships or relationships, and focus is usually on price as the key element. Yet the tendering process is expensive in time and money for the vendor. In industrial sectors such as electronics, tendering is usually performed on a given set of criteria in a relatively short time span. The general assumption is that the customer has a fairly clear description of its requirements, and that there are several suppliers who, with their capabilities clearly defined, are willing and able to provide the services.

Finally, Information Technology (IT) and Information Systems (IS) are increasingly used in supply chain management. Early applications focused on implementing sophisticated mathematical decision models for supplier selection. OSPAM (Optimal Selection of Partners in Agile Manufacturing) was probably one of the first attempts on extensive application of IT/IS in supplier selection, with particular emphasis on the Internet application (Minis et al, 1995). Vanwelkenhuysen (1998) has described a Tender Support Expert System for industrial centrifugal pumps. The system assists sales engineers to quickly generate and explore technically valid pump configurations as a response to customer requirements. Kroemker et al (1997) presented a concept of simultaneous bid preparation, and implemented a prototype infrastructure to support interdisciplinary co-operative bid preparation over a distributed heterogeneous system environment. These are only the beginning, and more work remains to be done in this direction.

This research has been induced with the aforementioned significance of the ESI in NPD approach, and thus the necessity for rigorous investigation. A web-based framework, called WeBid, is under development to promote and facilitate the actual implementation of ESI in NPD approach.

| New Product Development Process | | | |
Specification	Concept Design	Detailed Design	Production Design
• Establish specifications collaboratively • Avoid ambiguity and information distortion • Set technical targets • Articulate trade-offs • Identify early changes	• Key product and process technologies • Product architecture • Contribute key ideas/concepts/critical components • Participate in concept evaluation • Establish interfaces between product subsystems	• Selection of proprietary parts & components • "Black box" designed parts & components. • Tolerance design • Detail controlled parts & components • Prototype testing and demonstration • Design for Manufacturability • Material selection	• Make or buy decisions • Tooling & fixturing design • Equipment acquisition • Design for Manufacturability • Quality control & assurance • Raw materials
Supplier involvement			

Figure 11.1 Supplier involvement in new product development activities.

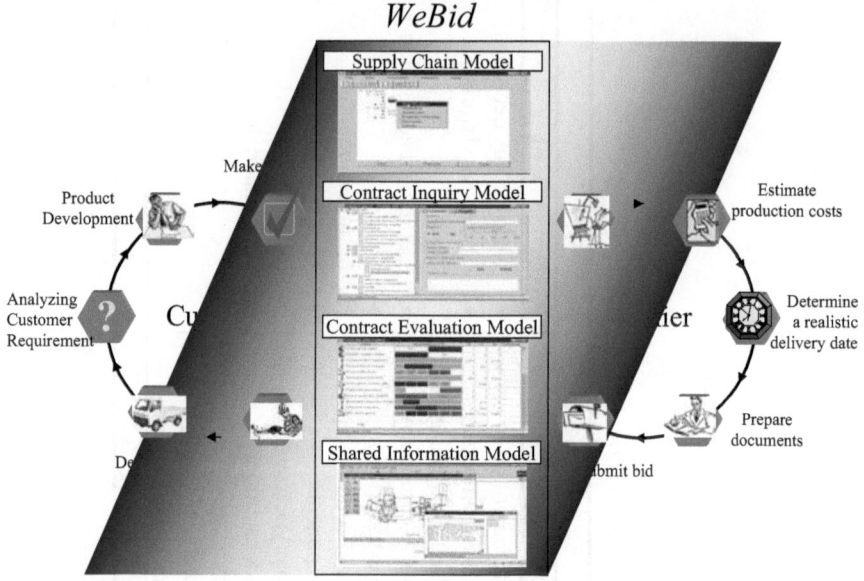

Figure 11.2 WeBid as an interface between the customer and supplier.

11.2.   THE *WEBID* FRAMEWORK

The scenario identified for this research concerns the initiation of a new product development project by a company. At certain stages of the development process, the company will have to consult its potential suppliers of raw materials, parts / components, tooling, equipment / machinery, and other types of consumables. Such consultation may take place as early as the design specification stage, or as late as the actual purchasing is due. Advance consultation with and participation of suppliers are termed as early supplier involvement. Contemporary research in the fields of concurrent engineering and supply chain management reveal that significant benefits can be achieved if suppliers are involved in new product development process as early as possible. However, recent investigation in manufacturing industries has also revealed that this approach is not widely practised in industries, and that its implementation has been a great challenge to researchers and practitioners.

11.2.1.   WeBid Overview

The aim of the present research is to develop an overall methodology to enlist better supplier involvement in new product development process, and to demonstrate the framework through a prototype web-based platform on the Internet/intranets using the web technology. Four key areas of study have been identified, namely, (1) to develop a product-oriented supply chain model that corresponds to the new product development process; (2) to develop a mechanism for the customer to invite potential suppliers to submit bids for manufacturing specific product components; (3) to develop a rigorous but pragmatic supplier selection methodology; and (4) to develop a mechanism to facilitate information sharing between the customer and suppliers.

Subsequent to an initial investigation, a prototyping system called WeBid has been developed. Figure 11.2 shows an overview of the WeBid system. On the left of Figure 11.2 are the main stages of the customer's new product development process, whereas its right side presents the main activities of the supplier's bid preparation process. Indeed, WeBid is seen primarily as an interface between these two processes.

11.2.2. WeBid Components

The middle of Figure 11.1 shows the four main modules of WeBid. These four modules also correspond to the four main activities of the general procedure of ESI proposed by Fine and Whitney (1996). The details of each activity will be presented in subsequent sections and the following gives a brief synopsis of each of them:

- The Supply Explorer. The Supply Explorer is the main and start-up component of WeBid. Other facilities are accessed through this module. The methodology underlying this Supply Explorer is a product-oriented supply chain model.
- The Bid Explorer. The Bid Explorer enables the customer to define their requirements and the potential suppliers to specify their supply capabilities. The methodology for this Bid Explorer is a hierarchical bid model shared by both the customer and suppliers.
- The Partnership Explorer. The Partnership Explorer enables the customer to evaluate and select potential suppliers based on their supply capabilities against the customer requirements. It uses a sophisticated, yet pragmatic and intuitive, quantitative partnership model. Four different types of numerical indices are used, namely, the satisfaction index, the flexibility index, the risk index, and the confidence index.
- The Share Explorer. The Share Explorer extends the scope of the system for the customer and suppliers in that sharing of information is not confined only to that of design but also design tasks, upon their reaching a mutual agreement on a project. The flow of information and work is administered through this component. This component is therefore of utmost importance and is worth substantial efforts of further investigation. Indeed, data interchange continues to be a great challenge to both researchers and practitioners alike. Due to the scope of coverage of this chapter, however, the design and development of this Share Explorer will not be discussed in detail. Nevertheless, the research

team's on-going investigation will report the latest findings in this respect in the near future.

Apart from these four main modules, WeBid provides a set of general utility facilities, e.g., registering with WeBid, partners search and query, browsing the "bidding" results, defining inquiry template, etc. Administrative facilities help maintain proper services for the WeBid service provider, such as managing the databases and web sites, etc. Registration facilities enable users to enter their details such as company name, address, telephone, username, password, etc. Login facilities allow users to obtain certain authorisation and to select a product development project from a list.

11.2.3.   Brokering Customer-Supplier Partnership

WeBid is a partnership broker for two types of users, namely, customer users and supplier users. Customer users are authorised to view but not manipulate the information of all the suppliers. Customer users can define a project for a new product, and can select one of his projects/products to create or modify the product hierarchy in the form of a tree structure at each stage of product development from conceptual design to production design. The Bid Explorer can be used to prepare design specifications for each item of the product tree. The customer may invite bids from relevant suppliers. After the deadline for submitting bids, the customer can calculate and evaluate partnership indices for each bidding supplier by calling the Partnership Explorer, and then select the best supplier(s) according to the optimal Partnership Indices for contract award on item production.

As regards supplier users, they can only view and cannot manipulate the customer information. Supplier users can manipulate their own information but not those of other suppliers. They can view invitations to their bids, or formulate plans for participation in the Supply Explorer. Once a decision is made, the user may launch the Bid Explorer to discuss the component design specifications with the customer and enter supply company's manufacturing capabilities to prepare the submission of a bid for the project.

11.3.   THE SUPPLY EXPLORER

Figure 11.3 shows that the Supply Explorer partially represents the supply web of a car front structure. The Supply Explorer is based on a Supply (Chain) Model developed particularly to support ESI in NPD. This supply model is comprehensible to both the customer and suppliers, and is consistent with the practices of their product development and bid preparation. The product model based on the Bill of Materials (BOM) seems to meet this requirement. For example, the MIT/Lehigh Fast/Flexible Manufacturing (Agile PathFinder) Project (Whitney et al, 1995) has extended the Bill of Materials into what they called the supply web. In addition, product models based on BOM have been used in bid preparation, as used in the European Pre-Bid project (Kroemker et al, 1997).

The present research extends the idea to formulate a model of product-oriented customer-supplier interface which is based on the concept of the supply web proposed by researchers involved in the Agile Pathfinder project (Whitney et al, 1995). The Bill of Materials for a car front structure shows a supply chain model of the Ford Explorer Front End. This map shows the parts, fixtures, and their respective vendors, and indicates that a large number of vendors can be available even for a small number of parts and fixtures. The bubbles containing "$, t, Q" inside highlight the major points where money and time are spent to obtain quality.

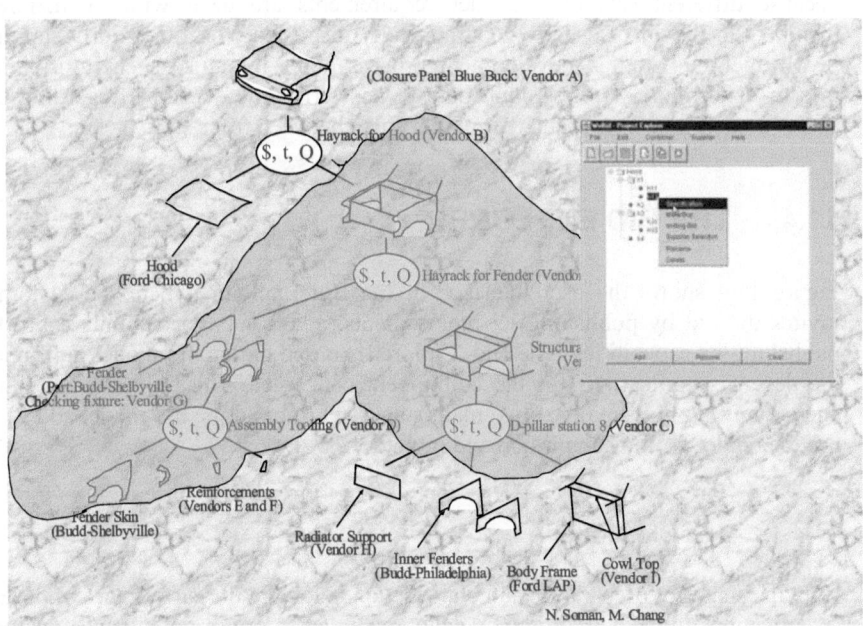

Figure 11.3 Product-oriented supply chain model.

Browsing through this Supply Chain model reveals an "as built" Bill of Materials (BOM) in indented form. At the top is the end product, such as a car or an aeroplane. The second level components are the major subsystems that are used to build the end product. Their suppliers are the first-tier suppliers with direct and close linkages to the end product manufacturer. First-tier suppliers, in turn, purchase components and materials from other suppliers and manage the "second tier" suppliers who, accordingly, manage a "third tier," and so on. In addition to component/materials supplies, this Supply Chain model also includes supplies of infrastructure technologies or equipment, and these are reflected in the nodes attached to process nodes for "Make" components or parts.

This Supply Chain Model constitutes a top-down product development approach. At the beginning, the product design starts with identifying the customer requirements of the end-product in the Supply Chain. Such customer requirements are then converted into product design specifications. This set of top-level requirements is broken down into sub-sets of requirements for the next lower level of subassemblies or subsystems. This process repeats recursively and the Supply Chain expands. New design activities occur and new members are added to the Chain. The requirements defined for the lower levels support the levels above, ideally in precisely defined ways, such as providing functions, physical supports, power, insulation, and so on. In reality, information on customer requirements and design solutions are easily lost or distorted in this Supply Chain, because different sets of customer requirements are dealt with by different members in the chain usually geographically dispersed over time. Early supplier involvement in product development process helps minimise such information loss or distortion. Suppliers contribute to answering the questions regarding the boundaries of subsystems and the relationships between these subsystems.

11.4.   THE BID EXPLORER

Figure 11.4 shows the user interface of the Bid Explorer, where the customer invites the bid by publishing its requirements, and a supplier submits a bid by publishing its capabilities. The Bid Explorer is based on the bid model built upon the design specifications for the component concerned. The process of inviting and submitting bids involves two main stages: 1) to build the bid model and 2) to use the model for bid submission and invitation.

11.4.1.   The Bid Model

Both the stages of preparation and clarification belong to the early stages of design specifications in the new product development/design process, which also happen to be an initial entry point for supplier involvement. With the convenience of having a set of design specifications at hand, the customer may start inviting bids from potential suppliers who are also able to prepare and submit a bid to the customer. In this connection, this research introduces the use of a model of design specifications as the common basis for the customer-supplier interface. Thus, the model of design specifications will be adopted as the bid model shared by both the customer and suppliers.

Since design specifications are the formal definition of customer requirements, and the model of customer requirements proposed by the authors (Lee, Huang and Mak, 1999) can therefore be extended and modified to form the model to represent design specifications. The items included in the bid/contract model are also called inquiry items or inquiries.

The main facility of this model is a hierarchical tree structure that is used to organise the specification items (also known as inquiries), as shown in the left-hand side of Figure 11.4. Various facilities are provided for users to build and elaborate design specifications, including the "list", "add", "edit", "delete", "copy" and "move" inquiries.

Inquiries may be in different types of values, and an inquiry type must be defined when it is added to the bid model. Thus, option buttons are provided on the right-hand side of Figure 11.4 (a), to supplement the general description and explanation of the inquiry. The four option buttons correspond to the following four types of inquiry values:

- Some inquiries belong to the type with continuous values. For example, when a customer requires the tolerance of a dimension to be within the Specification tolerance interval (+0.09 ~ -0.07) from $20^{+0.09}_{-0.07}$, these kinds of inquiries can be satisfied if the supplier's process capability can has a natural tolerance interval (+0.03 ~ -0.02) from $20^{+0.03}_{-0.02}$.

- Some inquiries are of the Boolean type, i.e., Yes or No. An example of such inquiries is "The company is ISO 9001 certified" and the answer to this inquiry is either yes or no.

- Some inquiries belong to the optional type with discrete values. For example, the customer requires several colours, red, yellow and blue, and the supplier is capable of providing beyond the requirements with green, red, yellow, blue and white.

- Some inquiries are qualitative and one quantifying method is to assign it a score with subjective ratings. An example of such inquiries is "After sales customer service" and the answer to this inquiry would be in the form of percentage from 0% to 100%. An added flexibility is that the score type can be converted to the range type in performing partnership evaluation.

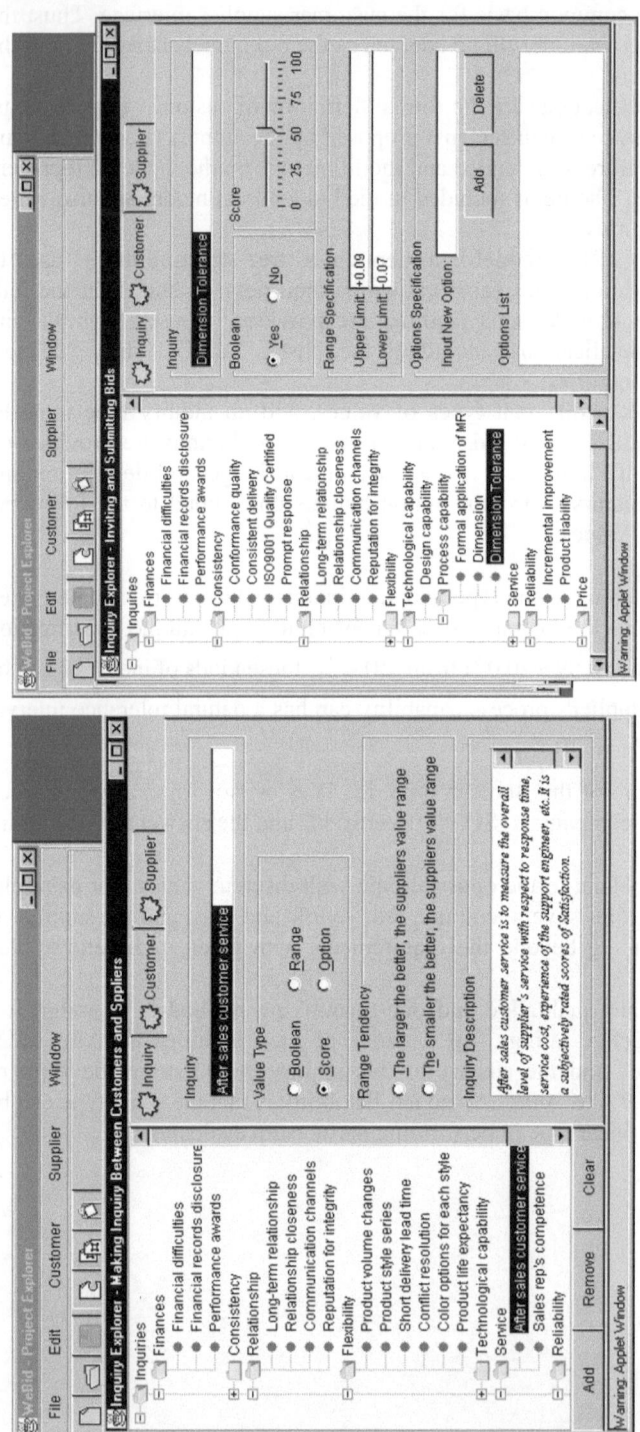

(a) Defining the bid inquiries

(b) Defining the values of the bid inquiries

Figure 11.4 User interface of the Bid Explorer.

11.4.2. Inviting and Submitting Bids

As soon as the customer and the supplier have reached a mutual agreement on the general items (inquiries) in the bid model, they may each start to define their requirements and capabilities quantitatively in their own contexts. The process of quantifying inquiries by the customer is called "inviting bids", and the quantified inquiries are called customer requirements. Similarly, the process of quantifying inquiries by the supplier is called "submitting bids", and the quantified inquiries are called supplier capabilities. Both processes can be called bidding, and the procedures for both processes are also identical. The value or value range for each inquiry in the bid model is specified using the facilities provided in the Bid Explorer shown in Figure 11.4 (b). The process is repeated until all the inquiries are considered.

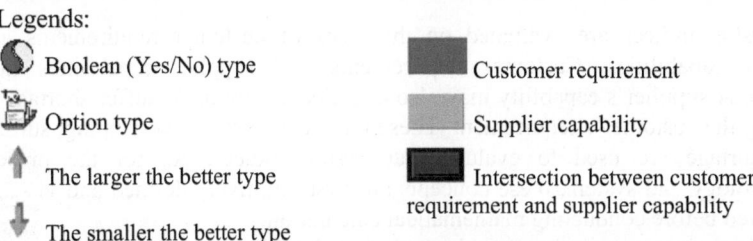

Partnership Explorer					
Inquiry	Visual Mapping of SI	SI	FI	RI	Comments
Financial difficulties	Yes / No	1	0	0	
ISO9001 Quality Certified	Yes / No	1	0	0	
Communication channels	Op0 Op1 Op2 Op3 Op4 Op5	0.400	0.330	0.600	
Product volume changes		1	0.57	0	
Product style series	Op0 Op1 Op2 Op3 Op4	0	1	1	
Short delivery lead time		0.83	0	0.170	
Color options for each style	Op0 Op1 Op2 Op3 Op4	1	0.400	0	
Product life expectance		0	1	1	
Formal application of MRPII	Yes / No	0	0	1	
Machinable Dimension Range		1	0.75	0	
Dimension Tolerance		1	0.69	0	
After sales customer service		0.875	0	0.125	
Overall		0.675	0.395	0.325	

Warning: Applet Window

Legends:

- Boolean (Yes/No) type
- Option type
- The larger the better type
- The smaller the better type

- Customer requirement
- Supplier capability
- Intersection between customer requirement and supplier capability

Figure 11.5 The Partnership Explorer.

11.5.   THE PARTNERSHIP EXPLORER

Figure 11.5 shows the main user interface of the Partnership Explorer, where the logic of the partnership model developed in this research is reflected. The partnership model includes four types of indexes as follows:

- Satisfaction Index (SI) is the measure of the level of satisfaction to customer requirement by a supplier capability. The larger the value of SI, the greater the potential for this pair of customer and supplier to become partners. Satisfaction index is based on the overlapping between the customer requirements and supplier capabilities.
- Flexibility Index (FI) is the measure of the extent to which a supplier capability exceeds a customer requirement and is based on the surplus of the supplier capabilities. The larger the value of *FI*, the higher the supplier's flexibility to cope with the changing customer's requirement.
- Risk Index (RI) is the measure of the possibility of a supplier's failure to meet a customer's requirement. The larger the value of *RI*, the more risky the partnership between them. Risk index is evaluated through the shortage of the supplier capabilities.
- Confidence Index (CI) is the measure of the supplier's reliability/competence in meeting the customer's requirements over a period of specified time. The higher the value of CI for a longer time, the more reliable the supplier is. A lengthened partnership may be considered instead of short-term competitive tendering. CI is evaluated through historical records of supplier performance, and is measured by selected indicators or inquiries.

These indices measure the extent of matching or incongruities between the customer requirements and the supplier capabilities, and therefore reflect the potential or risk of signing a project contract. The bidding party or parties with the best indices will be given further consideration of contract award.

11.5.1.   Individual Partnership Indices

Partnership indices are evaluated on the basis of customer requirements and supplier capabilities. Customer requirements and supplier capabilities may overlap. A supplier's capability may also experience surplus or suffer shortage in meeting the customer's requirement. These three concepts of overlapping, surplus and shortage are used to evaluate partnership indices. As for the present methodology and system, these concepts are first intuitively defined and visually illustrated before conducting mathematical calculation.

Multiple inquiries are involved in the bid model. The three types of partnership indices must therefore evaluate all the inquiries individually. Table 11.1 shows the formulas for calculating these three indices with respect to the different value types and value tendency of range values. In Table 11.1, the notations stand for values as follows: R_s - the set of supplier capability for a certain inquiry. R_c - the set of customer requirement for a certain inquiry. *OSI* - Overall Satisfaction Index. *OFI* - Overall Flexibility Index. *ORI* - Overall Risk Index. N - the total number of inquiries evaluated. SI_i - the SI for i th inquiry item, FI_i - the FI for i th inquiry item. and RI_i - the RI for i th inquiry item. Since there are three types of customer requirements (or supplier capabilities), R_s and R_c can be defined as two sets with either continuous value or discrete value, and indices must be evaluated differently for each type of bid inquiry value.

Table 11.1 Formulas to Calculate Partnership Indices

Value Type		SI	FI	RI
Range	R_s, the larger, the better	$\dfrac{R_s \cap R_c}{R_c}$	$\dfrac{R_s - R_s \cap R_c}{R_s}$	$\dfrac{R_c - R_s \cap R_c}{R_c}$
	R_s, the smaller, the better	$\dfrac{R_s \cap R_c}{R_s}$	$\dfrac{R_c - R_s \cap R_c}{R_c}$	$\dfrac{R_s - R_s \cap R_c}{R_s}$
Option		$\dfrac{R_s \cap R_c}{R_c}$	$\dfrac{R_s - R_s \cap R_c}{R_s}$	$\dfrac{R_c - R_s \cap R_c}{R_c}$
Boolean		$\begin{cases}1 & \text{if } R_s = R_c \\ 0\end{cases}$	0	$\begin{cases}1 & \text{if } R_s \neq R_c \\ 0\end{cases}$
Overall		$OSI = \dfrac{\displaystyle\sum_{i=1}^{N} SI_i}{N}$	$OFI = \dfrac{\displaystyle\sum_{i=1}^{N} FI_i}{N}$	$ORI = \dfrac{\displaystyle\sum_{i=1}^{N} RI_i}{N}$

Notes on set operators:	$A \cap B$ = Intersection of sets A and B.
	$A\text{-}B$ = Difference between set A and set B.

11.5.2. Uses of Confidence Indices

Confidence indices (CI's) are derived from the past records of performance measurements of a supplier (or a customer). The individual confidence index for a bid inquiry with a range of values can be defined as follows:

$$CI = \begin{cases} \dfrac{A}{C} & \text{if the inquiry item is of the larger the better type} \\[2ex] \dfrac{C}{A} & \text{if the inquiry item is of the smaller the better type} \end{cases}$$

Where "A" is the actual measured value or range of values of a bid inquiry when the supplier delivers the goods. "C" is the value or range of values of a bid inquiry that the supplier promises when a bid is submitted and subsequently awarded.

If the supplier's records over a certain period of time are available, an average may be taken to calculate the average confidence index for a bid inquiry. If more than one inquiry is used to evaluate confidence indices, then an average of such inquiries may be taken to calculate the overall confidence index (*OCI*) for the supplier.

There are two general uses of *CI* in partnership development. Firstly, the concept of confidence index is introduced for two purposes, as risk indices and, more importantly, as weighting factors to calculate the overall values of the other three types of indices such as satisfaction, flexibility.

Secondly, the overall confidence index (*OCI*) for a supplier can provide good indication in considering long-term partnership. If *OCI* is maintained at a high level over a certain period of time, then initiations can be made to negotiate a long-term collaborative partnership. Otherwise, this supplier must compete against any other potential suppliers for the contract in the future.

11.6.   THE SHARE EXPLORER

The Share Explorer extends the scope of the system for the customer and suppliers to share not only design information but also design tasks upon reaching a mutual agreement on a project. Both parties can upload or download design information, such as design drawings and process plans, from a central database. In addition, web-based design tools, such as Design for Manufacture and Assembly, are made available to both parties to carry out necessary analyses at the appropriate stages of the product development process.

However, the Share Explorer has not yet been fully demonstrated in this proof-of-the-concept WeBid implementation and is, instead, probably separated from WeBid to form another platform on its own. Nevertheless, the authors' research group has made some efforts. The following presents a brief summary of three aspects.

Firstly, two prototype systems have been developed for component cataloguing and concept design to support the suppliers to contribute their early concepts and off-the-shelf components during the customer's product design. Component and concept catalogues play an important role in supplier involvement in the customer's product development process. For those relatively standard and off-the-shelf components, selection can be made right after design specifications are completed, and suppliers can even contribute early concepts to the customer's product development project. In practice, only a subset of the specifications (usually functional) is used to select suitable components from the catalogues, and the key parameters of the chosen component are then used in the design.

Secondly, information such as product and process structures, key product and process characteristics, design drawings and virtual prototypes, etc., can and should be shared on the WWW between the customer and suppliers preceding and/or subsequent to their signing of contracts. In this connection, the concept of a synchronised Tele-Whiteboard has been investigated for both the customer and supplier to view the design drawings of common interest. Together with Tele-Chat, facilities are provided for members to make graphic or verbal comments on the drawing. This investigation has also brings about further research of previous studies on the application of Virtual Prototyping and Virtual Reality techniques on more sophisticated demonstrations of shared 3D solid models on the Internet using web browsers. In addition to geometric information, facilities will be provided to manipulate the parametric characteristics of products and processes.

Thirdly, the customer and the suppliers will need to share some of the design and development activities upon completion of contract agreement. In the past few years, the authors' research group has developed a number of web-based design tools. However, the target users for these tools were mainly for a closed community with all members from internal departments or units. Further investigations will focus on applying these tools in the context of supply chain management, and to investigate how such tools can be used to support distributed teamwork even from external participants. Appropriate modifications should be made so that they work in the customer-supplier environment. It is hoped that further guidelines for the design and development of web-based design tools can be compiled soon for future development.

11.7.   IMPLEMENTATION PERSPECTIVES

A prototype system called WeBid has been developed at the initial stage of investigation, with some sample screens discussed in the previous sections. The right side of Figure 11.6 shows the main facilities the WeBid system provides, whereas the left of Figure 11.6 shows a typical 3-tiered architecture of the web-based applications. The data source tier provides repositories of customer and supplier inquiries and the related information. The middle tier is a web server for the user to access and download WeBid facilities. Another type of middleware includes application servers that provide remote services to clients. It is not necessary for the web server and application servers to be identical. They may physically reside together in the same machine or in geographically disparate machines. The client tier includes client browsers with which the users connect to the WeBid web server. Indeed, all these three tiers can be separated physically, but they can be well-coordinated as long as they are connected to the Internet.

This section briefly discusses the various options that are adopted to design and develop WeBid and to deploy its main components.

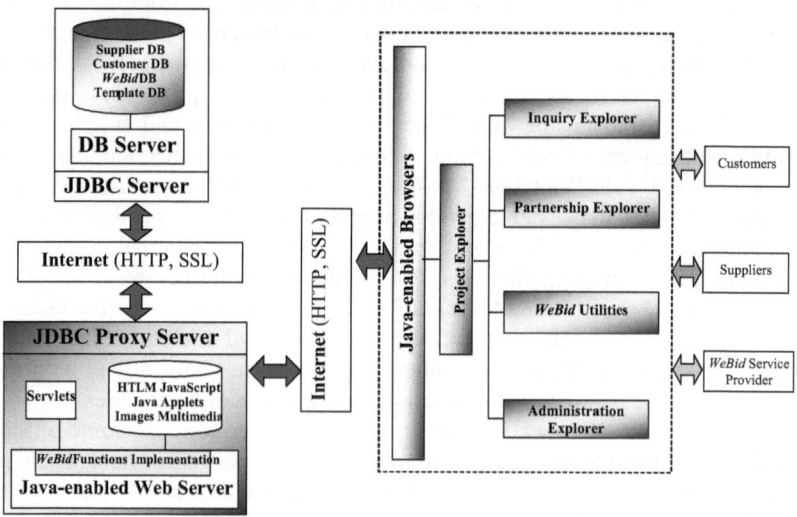

Figure 11.6 WeBid architecture.

11.7.1.   Design Considerations

The development, deployment and connection of WeBid facilities and database(s) are determined by their statuses. If WeBid is installed as a proprietary system maintained within a company, the WeBid database can be centralised and maintained by the owner company. If WeBid is installed as a third-party broker, two options are available. One is to maintain a central database by the service provider, a situation identical to that of a proprietary WeBid mentioned above. However, data security becomes an area of concern because proprietary information of user companies is kept by a third-party. Another alternative is to maintain a centralised database that contains only general information of user companies and their projects at the WeBid service provider while individual user companies maintain their own databases at remote sites. Proprietary information is maintained by the user companies themselves to avoid security problems.

11.7.2.   Distribution Considerations

In the client-server architecture, it is necessary to distribute the computation between the application server and the client, and to decide on the deployment of server and client components or objects. The two general methods of deploying web applications are "Fat" clients + "Thin" servers and "Thin" clients + "Fat" servers. As conveyed in their names, a "Fat" component is responsible for the majority of the computation, and a "Thin" component does little computation. In the initial development of this research, the client components under study are "fat" in the sense that they carry out most of the data processing. The server is only responsible for dealing with the connection to the remote databases.

The next subsection will demonstrate the application of Java to implement WeBid facilities, with several options of deployment available. In the first trial, WeBid was deployed as standalone Java applications without application servers. Later, attempts were made to deploy WeBid as Java applets with web servers and browsers, but without application servers. On-going efforts are being made to deploy WeBid as Java applets or standalone applications in combination with Java servlets that are associated with the web server. Eventually, it is anticipated that WeBid will be deployed as Java applets in combination with Java application servers.

11.7.3.   Implementation Considerations

There are two popular choices of implementing WeBid components, i.e., either ActiveX technology or Java-enabled technologies, since both technologies are capable of meeting the objectives of this project. Both Java and ActiveX technologies have evolved very rapidly and developments are still ongoing. It seems that JavaBeans are comparable to ActiveX components.

This WeBid project has chosen Java as the primary web and Internet programming environment to implement its components. The reason for this choice is simple. The authors' research group has applied the ActiveX technology extensively to develop various web applications in product design and manufacture, and significant insights have been gained with regard to this implementation environment. It is therefore desirable now to acquire more comprehensive experiences in Java programming, so as to obtain valuable information for comparison on the competence between Java-enabled implementations and ActiveX implementations.

11.7.4.   Database Deployment and Connection

In general, there are two commonly used methods to access the remote database on the Internet by a web application. One is the 2-tier model where a connection is built directly between the client and the remote database server, and a Java applet or application client talks directly to the database. This requires a JDBC driver to be installed and properly configured on the client machine, but users may be unaware of this requirement or are unwilling to do this before using the application or applet, despite its straightforwardness. There is no need to write server codes because everything is done by the client and the DBMS is the server.

The other method for remote database connection is using the 3-tier model to build a connection between the servlet or application server and the remote database server, and the client obtains the data from the servlet or application server. In the 3-tier model, all the database access operations are handled by the middle tier of servlets or application server objects. Clients invoke application server objects that, in turn, invoke the DBMS servers. Servlets usually reside in the application web server machine, while application server objects may be implemented in separate machine(s). In this model, the JDBC driver is installed and configured on the machine where the server object resides. It is therefore unnecessary to equip the client machines with the JDBC drivers.

11.8. SUMMARY

This chapter has outlined a framework that encourages suppliers to participate in the new product development process effectively and efficiently. The framework consists of 4 main mechanisms. The supply chain model (the Supply Explorer) is designed to be consistent with the new product development process. The Bid Explorer enables customers to invite and potential suppliers to submit bids for making specific components in a product. The Partnership Explorer corresponds to the supplier selection mechanism. Finally, the Share Explorer facilitates customers and suppliers to share information of common interest. An implementation of the framework on the World Wide Web (WWW or web) has been demonstrated. With acquisition of the initial experiences, the system has been re-designed to rationalise its workflow with new user interfaces.

12

WEB-BASED COLLABORATIVE PRODUCT DEFINITION

The process of product design and development process generally consists of four main stages: namely (1) planning and task clarification, (2) conceptual design, (3) embodiment design, and (4) detailing phases (Pahl and Beitz, 1984). In this research, the first two stages are considered as the early phase of product definition and the last two stages as the detail phase of product definition. As the names imply, the early product definition mainly concerns with establishing key characteristics of a product under development while the detail product definition deals with concrete descriptions of the subject product.

This chapter will focus on development of software support for the early product definition on the Internet and World Wide Web. There are a number of reasons for this focus. First, the early product definition stage is very crucial. Because after this stage rework or redesign is rather expensive, and sometimes even impossible (Pahl and Beitz, 1984).

Second, the early product definition stage is the most innovative one. In addition to its abstract complexity, it requires considerable amount of experience and skills from designers and development teams. Therefore, it is necessary to develop a tool that can adequately support designers in the early design process (Hague and Bendiab, 1998).

Third, the early product definition stage only receives little support and help from recent design tools because the abstract process is difficult to model during the conceptualisation process (Chakrabarti and Bligh, 1996). Majority of the design systems, such as CAD and material selection software, support and facilitate the detail product definition stage only. Thus, the success of the early product definition still heavily relies on designers' knowledge and experience.

Finally, the early product definition stage involves people from different disciplines at different locations. It is therefore important to develop design tools, which can facilitate the collaborative nature.

Web applications are one of the solutions that are gaining increasing popularity in these few years (Erkes *et al*, 1996). By implementing design tools on the Internet, geographically distributed design teams can work together and different people can participate in the design process through the web and give out their ideas. This chapter will present an effort made in developing an Internet-enabled support, called ProDefine, for the early product definition collaboratively through a distributed project team. Section 12.1 is an overview of system methodology, architecture, and other components of the early product definition. Sections 12.2, 12.3, 12.4, and 12.5 will discuss the details of Concept Explorer, Requirement Explorer, Generation Explorer, and Evaluation Explorer respectively. At the end, Section 12.6 will try to draw some implications.

12.1. FRAMEWORK FOR EARLY PRODUCT DEFINITION

This research aims to develop a generic framework for the early product definition at the concept stage of product design and development cycle, and also to demonstrate the methodology through a web-based system.

12.1.1. Methodology for Early Product Definition

The essence of the Early Product Definition methodology adopted in this chapter is developed based on the Morphological Chart Analysis method – a well-known formal design tool. The method was first introduced by Norris (1963) and has been adopted and included in a number of major textbooks on Engineering Design as one of the most effective methods for concept design (Pahl and Beitz, 1984; French, 1985; Cross, 1994; Ulrich and Eppinger 1995).

Although a few minor variations among these concept design models are inevitable, their basic elements are three major stages, including (functional) requirement analysis, concept generation, and concept evaluation. Requirement Analysis stage mainly concerns itself with establishing functional and other design requirements of the product, which is under development. Concept Generation stage identifies potential solutions to meet the functional requirements established at the previous stage. A Morphological Generation Chart shows feasible concepts for each requirement. Solutions are defined by choosing different combinations of concepts in this chart. Concept Evaluation stage carries out comparative analysis of potential solutions against different criteria. Three components are involved in this stage. A list of criteria is firstly established. Each solution can be evaluated by the other two components and they are Quality Function Deployment (QFD) and

Morphological Evaluation Chart. QFD is used to verify whether the solution meets all the customer requirements while Morphological Evaluation Chart uses the identified criteria to compare different solutions against the identified criteria.

They are configured to form the overall methodology of early product definition as shown in Figure 12.1. The outcome of this early product definition phase is a preliminary layout design that reflects main working principles and features of a product. During this process, substantial amount of information is needed and as well as being generated.

12.1.2. *ProDefine* System Architecture

After initial efforts, a prototype system, called ProDefine, has been developed The ProDefine architecture as shown in Figure 12.2. It fully corresponds to the overall methodology shown in Figure 12.1. The three key stages in the early product definition are supported by the three key ProDefine components, which are namely Requirement Analysis Explorer, Concept Generation Explorer, and Concept Evaluation Explorer.

In addition to the three key components, there are also two design data reservoirs involved to support the early product definition. They are the Concept Knowledge Base and the Design Reservoir. These two data reservoirs are the data storage areas for the current and past design knowledge and solutions. Concept Knowledge Base is a storage space for past design concepts, providing new design concepts for new products during the process of concept formation. Design Reservoir is used to store up design decisions, such as functional requirements of the Requirement Analysis stage, solutions derived in the Concept Generation Stage and evaluation results from the Concept Evaluation stage.

Concept Explorer (browser/editor) allows a geographically distributed team to design and develop the structure and content of the Concept Knowledge Base and there are three possible types of information flow: between the Concept Explorer and the Concept Base; between the Concept Generation Explorer and the Concept Base, and between the Concept Evaluation Explorer and the Concept Base. There is no direct flow of information between these three modules. In contrast, the flow of access is more liberal in the sense that any of them can be accessed directly, either through the home page or from other modules. Hyperlinks are provided in each module to go to the other two modules.

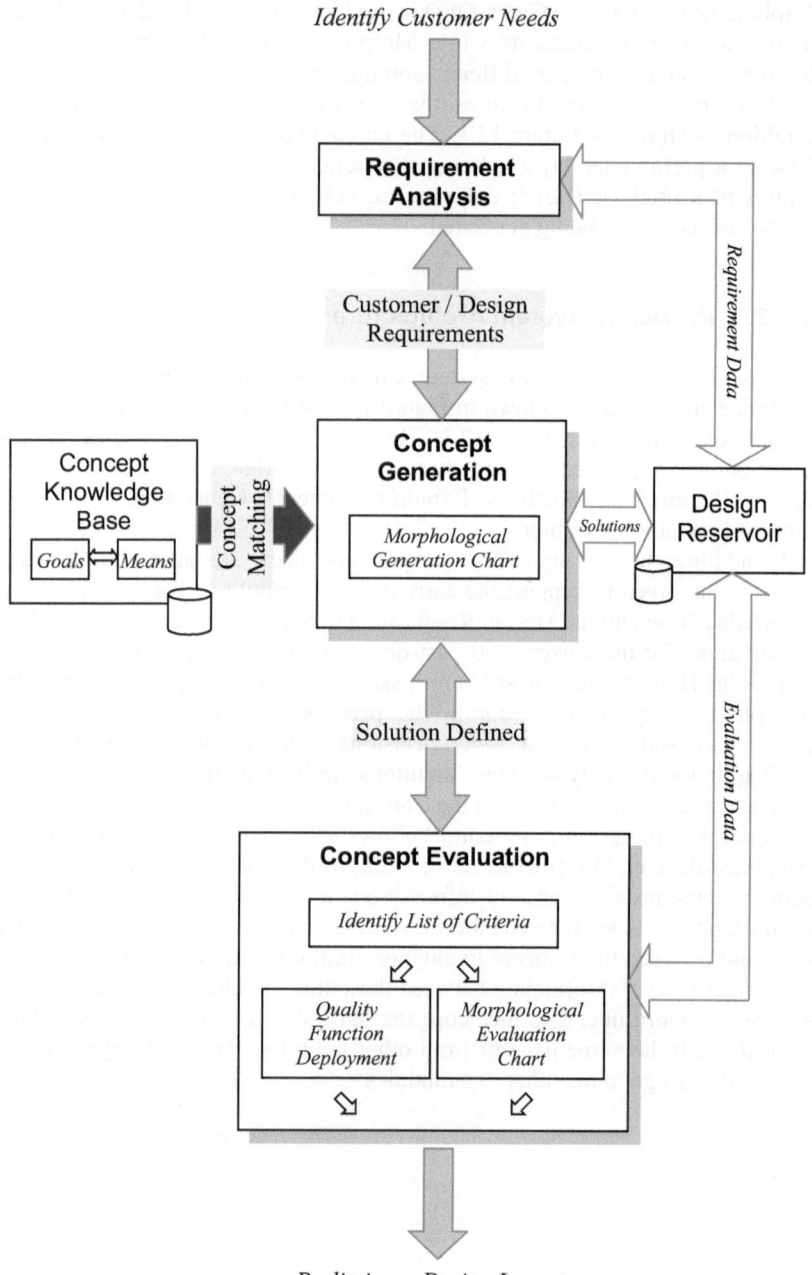

Figure 12.1 Overall methodology of early product definition

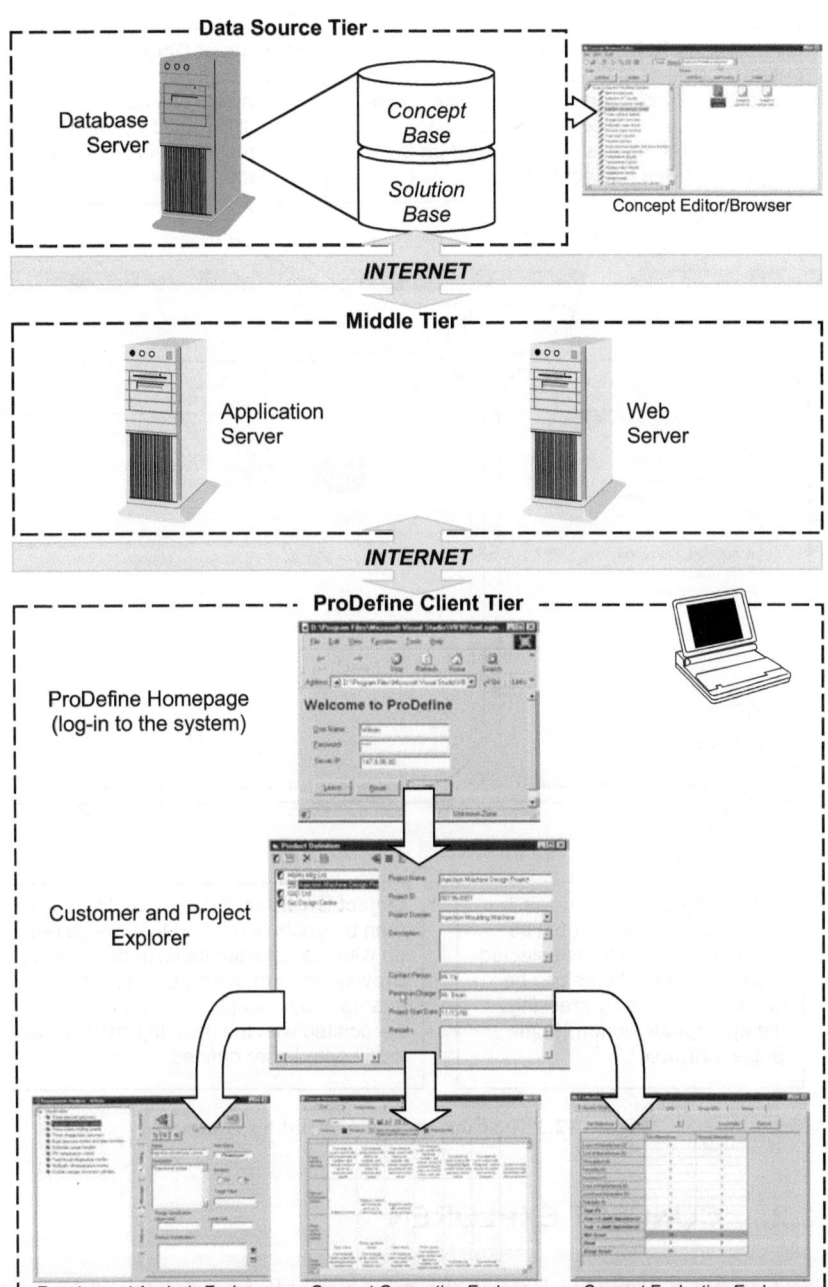

Figure 12.2 Overall System Architecture of ProDefine.

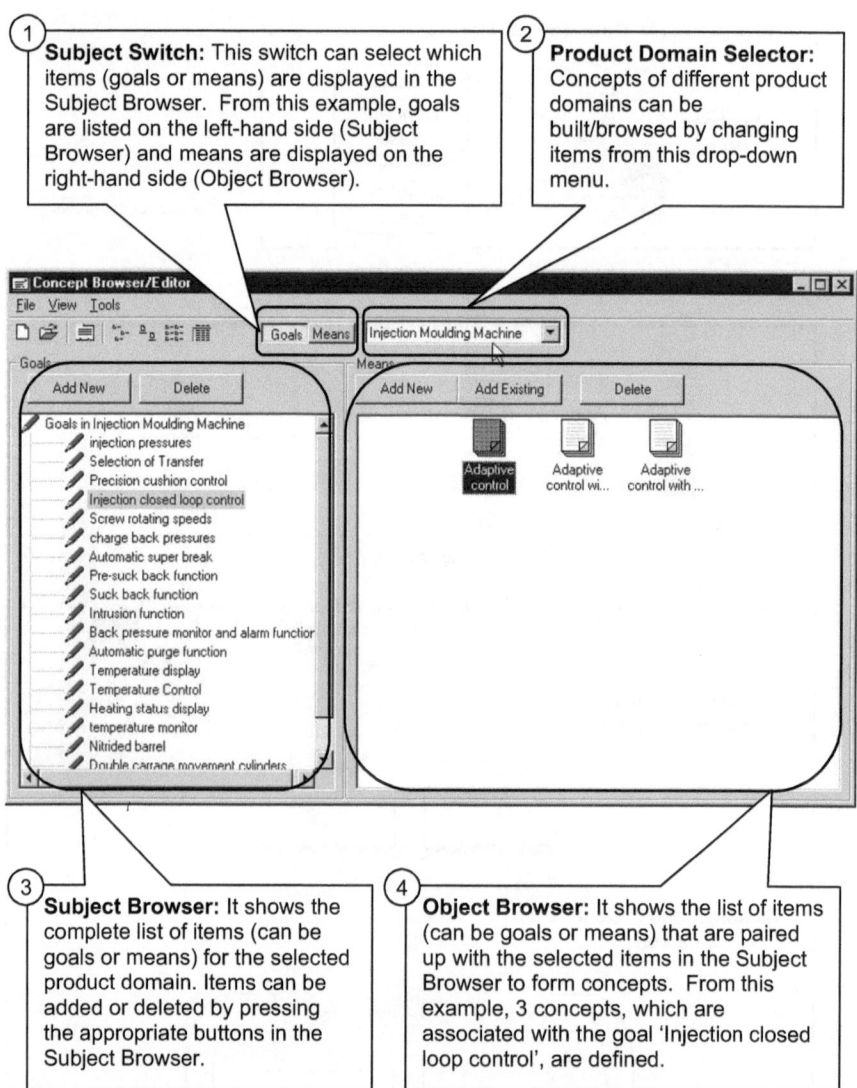

Subject Switch: This switch can select which items (goals or means) are displayed in the Subject Browser. From this example, goals are listed on the left-hand side (Subject Browser) and means are displayed on the right-hand side (Object Browser).

Product Domain Selector: Concepts of different product domains can be built/browsed by changing items from this drop-down menu.

Subject Browser: It shows the complete list of items (can be goals or means) for the selected product domain. Items can be added or deleted by pressing the appropriate buttons in the Subject Browser.

Object Browser: It shows the list of items (can be goals or means) that are paired up with the selected items in the Subject Browser to form concepts. From this example, 3 concepts, which are associated with the goal 'Injection closed loop control', are defined.

Figure 12.3 The Concept Explorer (Editor/Browser).

12.2. CONCEPT EXPLORER

The Concept Knowledge Base is the most valuable resource for early product definition in the ProDefine and the Concept Explorer is the only facility that allows the user to input enter new concepts into the Concept Base and/or look at its existing concepts contents.

12.2.1. Editing/Browsing Concepts

Figure 12.3 shows the layout of the Concept Explorer. It provides various basic functions to manipulate the Concept Base, including adding new concepts, deleting and updating existing concepts. The Concept Editor/Browser aims at providing an easy-to-use application for users to build up the Concept Base. Four main components are involved. They are Product Domain Selector, Subject Switch, Subject Browser and Object Browser. Product Domain Selector (Item 2 in Figure 12.3) is a drop-down menu that consists of a list of available product domains. It allows users to choose which product domain to work on. Logically, each product domain should have a separate Concept Base. Therefore, the editor will switch to the concept base of another product domain whenever the selection in the product domain selector is changed.

Subject Switch (Item 1 in Figure 12.3) is responsible for defining layout representation of the editor. It determines which items (goals or means) should be listed in the Subject Browser. It consists of two buttons: 'Goals' and 'Means'. When the button is pressed, the corresponding items will be listed on the Subject Browser and the remaining ones will be shown on the Object Browser.

The relationship of "Goals" and "Means" is displayed in the Subject Browser and the Object Browser. Subject Browser (Item 3 in Figure 12.3) is on the left-hand side of the editor. It shows a list of either goals or means, which is determined by the Subject Switch. On the contrast, Object Browser (Item 4 in Figure 12.3) is on the right-hand side of the editor. It shows defined concepts that respectively correspond to the selected items of the Subject Browser. In other words, each item in the Object Browser can be paired up with the selected item in the Subject Browser to form a concept.

Take Figure 12.3 as an example, 'Goals' is selected in the Subject Switch. This indicates that the goal of injection moulding machine is listed on the Subject Browser while the means are displayed on the Object Browser. There are three means displayed on the Object Browser. This suggests that three means are associated with the goal 'Injection closed loop control', resulting in three defined concepts in the Concept Base.

As what we have observed, the layout of Concept Editor/Browser allows a flexible representation of the relationship of "Goals" and "Means". Users are free to browse or search, and see which goals are associated to a means or which means are associated to a goal. This is accomplished through the introduction of the Subject Switch.

12.2.2. Concept Representation and Implementation

The design of the Concept Base is based on the notion of concepts. A concept can be described as "Goal is Achieve(d) by Means" or "Means Achieve(s) Goal". 'Goals' are the functional goals of the product domains and 'means' are the feasible ways to achieve these goals.

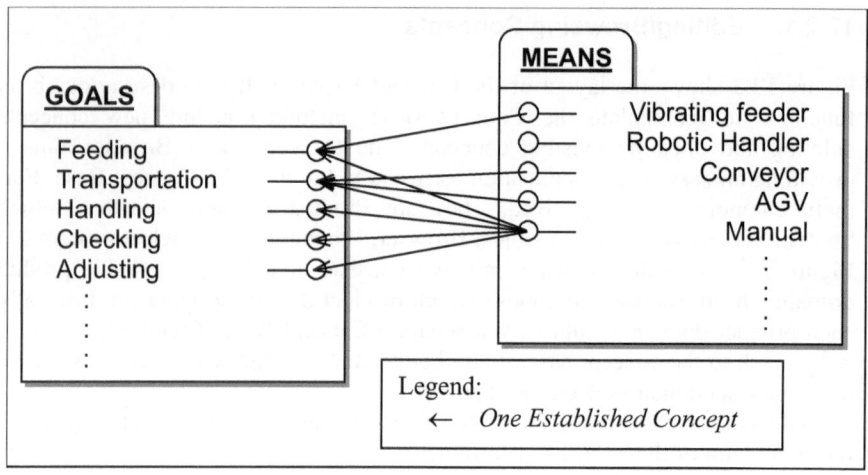

(a) A concept as a pair of Goal and Means.

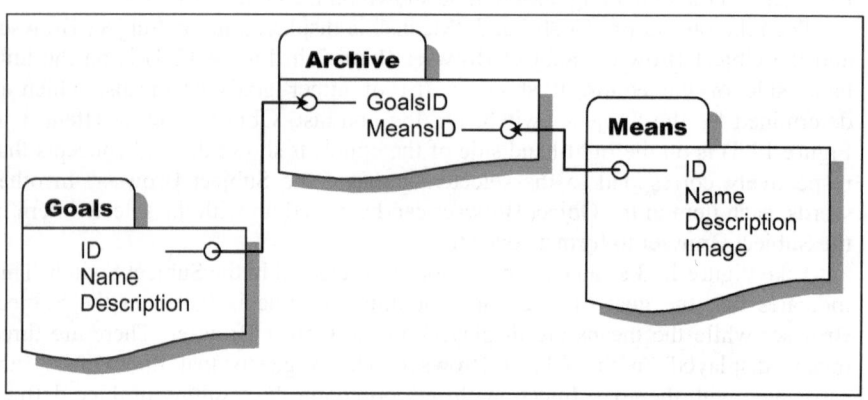

(b) Relationship of Goals, Means and Archive Tables in the Concept Base.

Figure 12.4 Concept representation and implementation.

A case study has been carried out to illustrate the process of building a sample Concept Base. This case study is based on the work done by Rampersad (1994). The domain of the case study is the design of assembly systems for electrical plugs. "Goals" are the desired functions of an assembly system, including feeding, transportation, and handling, etc. Means are often used to achieve these functions. At present, about forty means, such as vibrating feeder, robotic handler, and conveyor are collected in the Concept Base to achieve eight different assembly functions. By incorporating more assembly operations and alternative means, the Concept Base can be further expanded.

The relationships of 'goals' and 'means' are described as concepts, as shown in Figure 12.4(a). As a result, a concept can be depicted as "Goal is achieved by Means" or "Means achieve Goals". For example, the goal "Transport" in the context of assembly system design can be achieved by means of "Manual transport", "Conveyor transport", and "AGV transport". Each combination is a concept. As a result, there are three concepts involved in this example and they are "Transport is achieved by Manual transport"; "Transport is achieved by Conveyor transport"; and "Transport is achieved by AGV transport".

The mapping of 'goals' and 'means' can be one-to-one or one-to-many, even many to many. A goal can be achieved by a number of means and a means can also achieve a number of goals. Each mapping of a goal and a means is referred as a concept. Back to the previous example, the means "Manual transport" can achieve different goals, like "Transport", "Feeding", "Checking", "Adjusting" and "Handling", etc. At the same time, the goal "Transportation" can be achieved by different means, like "Conveyor", "AGV" and "Manual Transportation", etc. This mapping is illustrated clearly in Figure 12.4.

The Concept Base is implemented by using a relational database containing descriptions about individual concepts and their details, as shown in Figure 12.4(b). Following the notion of goals and means. The Concept Base primarily consists of three tables: the "Goals" table, the "Means" table, and the "Archive" table. The 'Archive' table speaks for the relationship of "Goals" and "Means". Each record in the "Archive" table represents one concept. Each record is a pair of one goal and one means from the 'Goals' table and the 'Means' table respectively.

12.2.3. Collection of Concepts

Relevant concepts must be first collected into the Concept Base before it can be used for generating solutions to a design problem. The process of collecting concept collection is complicated and effort-consuming. The first issue is about the source where the concepts are identified. The following are some possible sources:

- *Books*. Reference books are one major source of concepts. Design handbooks for the product domain can provide a list of goals and functional alternatives to achieve the same goal.
- *Past Designs*. Concepts can be identified from past designs, such as part list, part drawing, and past design notes etc. Moreover, more than one design variation or model from the product domain should be available for extracting different means for the same goal.
- *Past Experience*. Past experience of designers can contribute to the Concept Base. The designers should be well knowledgeable and well experienced so that they can specify the goals and functional means. Moreover, they should be informed with the most up-to-date information concerning the product domain.
- *Suppliers' catalogues*. The manufacturers' suppliers can contribute to the Concept Base by providing information on machine components or

mechanism, whenever new technology is evolved. Certain concepts can also be extracted from suppliers' catalogues.

After identification, concepts must be properly processed in terms of their format and contents so that they can be inputted into the Concept Base. This conversion process may appear to be a simple task. However, it is very complicated and tedious even with the simple concepts used in our illustrative case study. It requires further in-depth investigation.

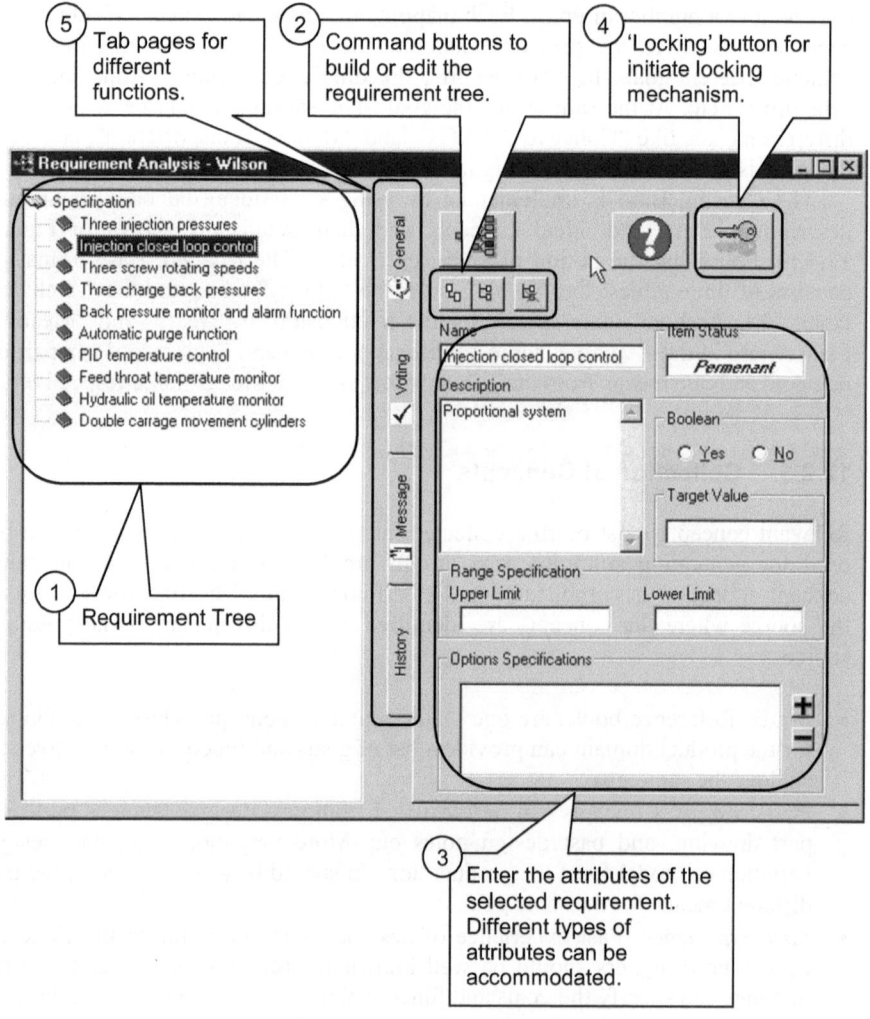

Figure 12.5 The Requirement Explorer.

12.3.   REQUIREMENT EXPLORER

Requirement Analysis is an early stage in concept design. It is a critical stage for designers to detect and capture customer needs, and then to convert them into design specification (Tseng and Jiao, 1997). It has important implications for new product's success or failure. Therefore, it is necessary to define a requirement list that can truly reflect both customer needs and market needs. The Requirement Explorer is responsible for this purpose.

12.3.1.   Requirement Representation

Figure 12.5 presents an overview of the Requirement Explorer. It includes two main parts: a tree structure representing design requirements on the left-hand side and a form for specifying details of requirements on the right-hand sides, respectively. The requirement list in the Requirement (Analysis) Explorer is organised in hierarchical format. Usually, the higher-level requirements will record the abstract customer requirements. The lower-level requirements are the ways to achieve the higher-level requirements. By expanding the branches, concrete functional requirements can be reached at the bottom level.

The choice of using hierarchical structure is made after reviewing various methods for functional (requirement) analysis (Cross, 1994), such as functional flow charts, functional block diagrams, objective tree analysis, and functional logic diagrams, etc. Each has its advantages and disadvantages and none is considered perfect (Sturges et al, 1993). The hierarchical method adopted here is based on function logic diagram analysis or objective tree analysis. This was a formal approach established by Bytheway (1971) in order to facilitate Value Analysis (Miles, 1965), the resulting technique was named FAST (Functional Analysis Systematic Technique) and the resulting diagrams are called FAST diagrams.

A tree-structured requirement list has several advantages. Firstly, trained and experienced engineers are familiar with the FAST method and diagrams because they have been adopted in major textbooks and Value Analysis is widely practised in industries. In addition, the method is relatively simple because functions are linked together to form a hierarchical tree, in which only why/how relationships are represented.

Secondly, a requirement tree can provide a means of 'thought ordering' and most people can understand it easily. In order to develop a requirement tree, the top-level requirement has to be identified and defined first. It is then broken down into sub-requirements by asking 'what do we mean by this' (Wright, 1998). As a result, a main objective is divided into smaller objectives that are easier to achieve.

Thirdly, this is a clear way of representing complicated ideas, because a large number of requirement items are usually involved in a design project. If a flat list is being used, then it will be very difficult to manipulate and work with in the later design stage. Therefore, it is necessary to categorise requirements into a hierarchically ordered list.

Finally, a requirement tree can support collaborative design more effectively. Since the requirements are categorised into sub items, a design team can work out the structure of the requirement list together first, then work collaboratively by allocating members to work with different categories. Afterwards, each team member will be responsible for building requirements to one or more categories.

12.3.2.   Compiling Design Requirements

The use of this explorer is straightforward. However, the user must be aware of several issues. Firstly, the user should define his/her own scope and meaning of "design requirements". Different people tend to use the term for different their own purposes. The meaning of Design Requirements should be taken liberally to cover design objectives, user needs or product purposes. Regardless of the name, it is the concrete aims that the new designs trying to achieve (Cross, 1989). The scope of the meaning depends on the focus of the design project. Some user may focus on Functional Requirements only (which is only a sub-set of design requirements). Some user may try to extend their efforts to look at cover all the relevant aspects, such as product performance, selling prices, and reliability, etc. However, the method incorporated into the Requirement Explorer applies to all these situations.

The second issue is related to the classification of design requirements. Sub-requirements can only refer to one 'parent' in the hierarchical representation. This indicates that a sub-requirement belongs to one category only. However, there are some requirements that belong to more than one category and they cannot be represented by this method. For example, the design requirement 'Low Risk of damage to workpiece or tool' can be classified under different requirements: 'Machine must be safe' or 'Low Defect Rate'. Obviously, the tree representation is incapable of dealing with this kind of situation. People will have different opinions on the structure of the requirement list (Cross, 1989; Wright, 1998). This is because the requirements can be structured by using different form of classification, which can explain the same idea effectively. This may probably lead to conflicts when some team members insist other to follow their decisions. Nevertheless, according to Suh's Design Axioms, functional requirements should be better independent of each other. In other words, multiple relationships between functional requirements are undesirable.

Thirdly, the usual procedure of defining design requirements using the Requirement Explorer follows a top-down approach, namely proceeding from the highest level to the lowest bottom level. However, a bottom-up approach is also possible with this Functional Analyser. A flat list of functions is first defined. The drag and drop facility provided in this function editor is used to build the tree. However, it is recommended to follow the top-down approach in the context of morphological chart analysis. The bottom-up approach should be carried out using a full functional analysis that is beyond the scope of this chapter and will be dealt with separately in a later stage.

The final issue is how to compile a list of customer requirements and from where to start. This is outside the scope of this research, too. However, designers should have a thorough understanding of customers and users' needs, competitor information, technology risks, (Bacon *et. al.*, 1994) and all other parameters that may affect new product generation. This background knowledge is the input of the product definition. It can ensure that a reasonable and feasible requirement list is being built. Bacon *et. al.* (1994) showed that successful product design and development teams paid more attention to information competing products. The prospective product offerings of competitors can give insights to successful product design teams. Together with assessment of customers' needs, it gives more understanding of the current market status and how other companies meet their requirements. In addition, an analysis of availability and reliability of technologies used in a product or its manufacture is another important input into requirement analysis (Bacon *et. al.*, 1994). These can ensure a feasible design will be developed during the early product definition. Moreover, these can help design teams to estimate the production cost more accurately.

12.4. CONCEPT GENERATION EXPLORER

Concept generation is a divergent and creative stage of conceptual design. Its purpose is to conceive as many solutions as possible to sub-functions (requirements), and to produce as many feasible solution alternatives to the overall goal as possible. The input to this stage is the hierarchical tree of (functional) requirements as established in the preceding stage and the output from this stage will be a set of solution alternatives to be forwarded to the next stage for in-depth evaluation. The conversion from the input to the output will make full use of the Concept Base and the user's skills and expertise in the application domain.

12.4.1. Morphological Generation Chart

The basic mechanism used in the Concept Generator is a morphological generation chart. Figure 12.6 shows a sample Web page of the concept generation matrix. The vertical dimension of the matrix represents a list of basic functional goals, namely the functional requirements at the lowest level. This list is therefore flat, no longer hierarchical. The horizontal dimension of the matrix represents all the conceivable means that are relevant to achieve the corresponding basic functional goals. Different goals may have different numbers of conceivable means. Hence, the numbers of cells of the rows in the chart may be different. Based on this morphological chart, the designer is able to make further analysis and choices regarding which means is more applicable.

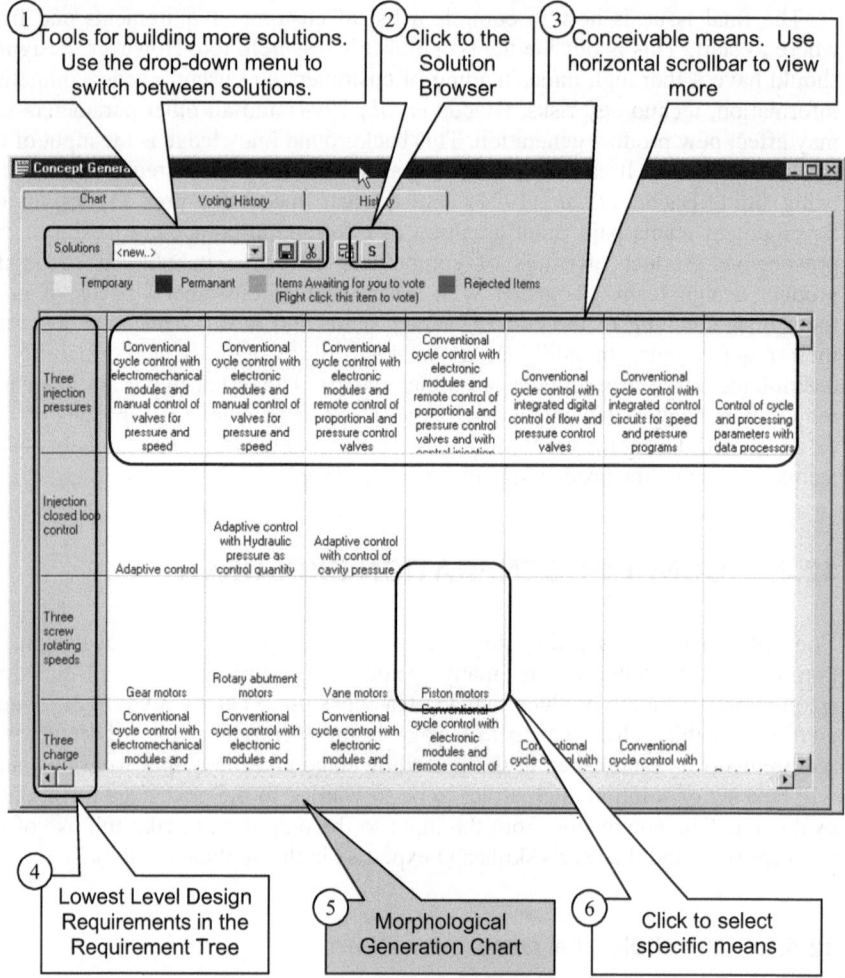

Figure 12.6 The Concept Generation Explorer.

The general steps involved in generating and using the morphological chart for developing design concepts are as follows:

- Generating conceivable concepts from the concept knowledge base according to the design requirements.
- Defining feasible solutions that seem to meet the design requirements individually and in overall terms.

12.4.2.   Generating Conceivable Concepts

Concepts are first conceived for the design requirements individually based on the concept knowledge base. All the conceived concepts for a particular requirement are presented along the corresponding row. This conception process seems to be straightforward and automatic to the user. However, one question still remains, i.e. how do we determine if a concept is conceivable in relation to a particular requirement. The conception is based on a matching algorithm between the descriptions of design requirements and descriptions of concepts. Such mappings are usually difficult to determine and deserve further in-depth investigation.

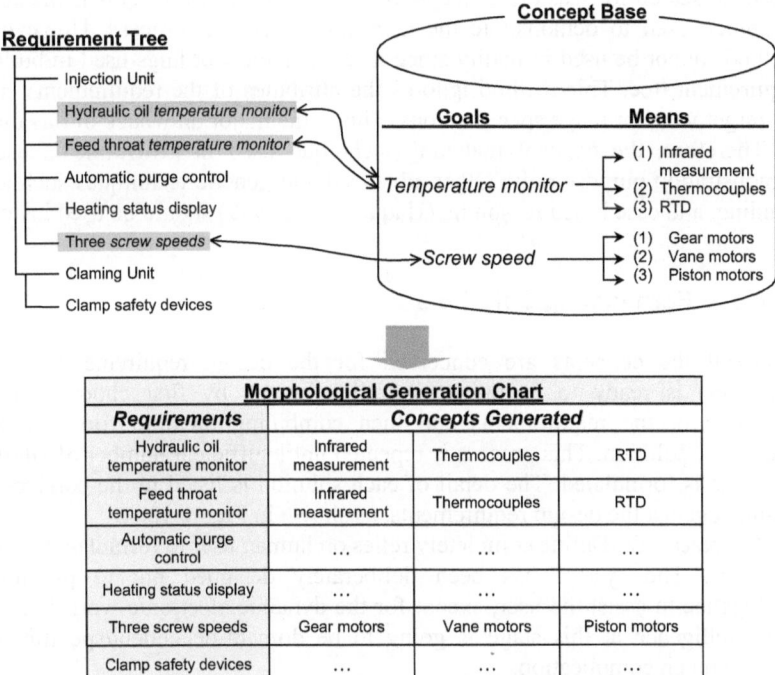

Figure 12.7 The illustrative example of concept matching.

It should be noted that concept matching is not the main focus of this research. The prototype ProDefine system only provides some rudimentary criteria for retrieving conceivable means. The concept matching technique involves a direct matching of design requirements to the goals in the Concept Base. Figure 12.7 shows an illustrative example of how concept matching takes place. On the right-hand side of Figure 12.7 is the Concept Base, showing the relationship of goals and means. The requirement tree for that product domain is shown on the left-hand side. The arrows between them represent that the goal in the Concept Base is matched to the requirement item.

The concepts are matched when the names of the goals appears in the names of the requirement items. For example, the phrase "*temperature monitor*" appears in the name of a requirement "*Hydraulic oil temperature monitor*". The Association of the means and the goals "*Temperature monitor*" (*Infrared measurement, Thermocouples* and *RTD*) will be the suggested concepts for the requirement "*Hydraulic oil temperature monitor*". The same concept will be matched to a requirement "*Feed throat temperature monitor*" since the phrase "*temperature monitor*" appears on the name of the requirement.

This concept matching process is achieved by applying the SQL keyword 'Like' to search for feasible concepts from the Concept Base. This is the simplest technique used to demonstrate the Concept Generation Process. However, this method cannot be used in reality since it restricts the wordings used in building a requirement tree. This method ignores the attributes of the requirements such as the target value or range specifications. This is the major drawback of this method.

Therefore, the concept matching technique used in ProDefine is used for demonstration purpose only. Other advanced and generic techniques such as data minding, and case based reasoning (Haque *et. al.*, 1999) should be used instead.

12.4.3.   Formulating Solutions

Once all the concepts are conceived for the design requirements, the user (designer) is ready to formulate overall solutions by first choosing feasible concepts for the requirements and then combining them to form an overall alternative solution. This process is repeated until sufficient number of alternative solutions is formulated. The detail of each solution is listed in the corresponding column against the design requirements, as shown in Figure 12.8.

At present, ProDefine completely relies on human user to formulate the overall solutions. The system has been deliberately designed not to provide any intelligence to assist the user, except for the dynamic electronic worksheet. Also, any intelligence at this stage is going to be domain-dependent/specific, which leads to much complication.

When formulating solution alternatives, the user must keep the following two issues in mind. The first issue is the criteria for determining if a conceivable concept is feasible in a particular application. As explained above, the ProDefine system at present does not provide any suggestions about the feasibility criteria, and the human user is fully responsible for making such judgements.

The second issue is related to the combination of individual means into overall solutions. Mathematically, a vast number of combinations are possible. Assume that there are 10 sub-functions and three means are short-listed for each sub-function. Consequently, there are 3^{10} (=59049) possibilities of combinations. This number is obviously too large to manage. Therefore, certain domain-specific rules have to be applied to manipulate the production of combinations. At present, the system solely replies on the human user to control the combinatorial explosion.

Design Requirements

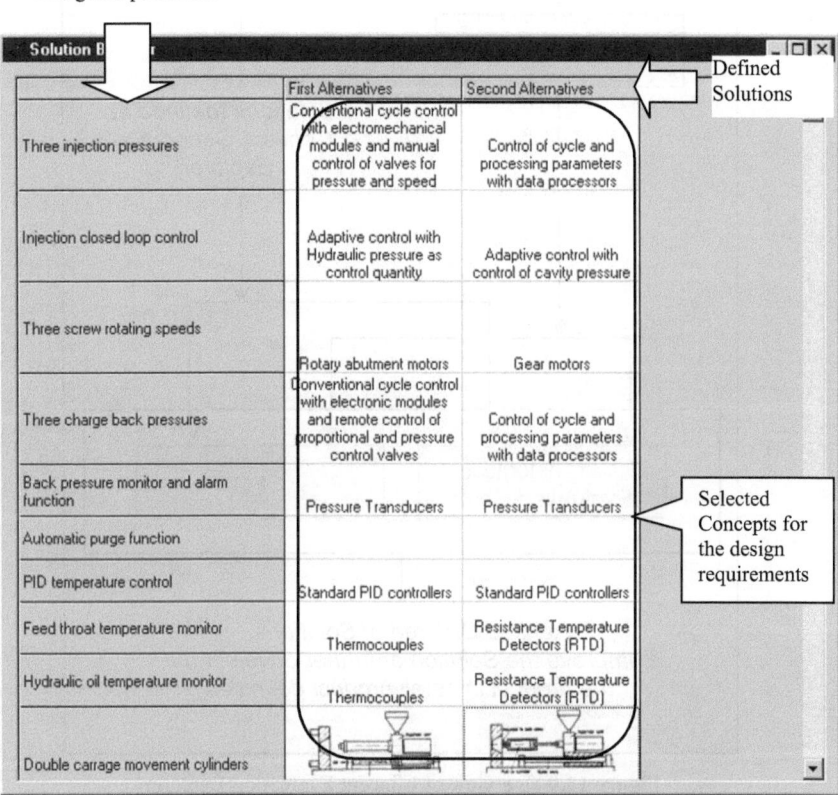

Figure 12.8 Solution Browser.

12.5. CONCEPT EVULATION EXPLORER

Concept evaluation is a convergent stage of conceptual design. Figure 12.9 gives an overview of the concept evaluation explorer. The purpose is to narrow down the number of the feasible solution alternatives for further investigation. The input to this stage is a list of solution alternatives obtained from the previous stage of (Concept Generation), and the output of from this stage is a rank order of the inputted solution alternatives to be forwarded for further development. The conversion from the input to the output is based on a set of criteria against which alternative solutions are being evaluated. In addition, to the evaluation of alternative solutions, ProDefine also offers another method based on QFD for evaluating individual constituent components of the solution. The human user plays a key role in providing performance measurements. This is rather time-consuming and data intensive to assess alternative solutions against individual criteria one by one.

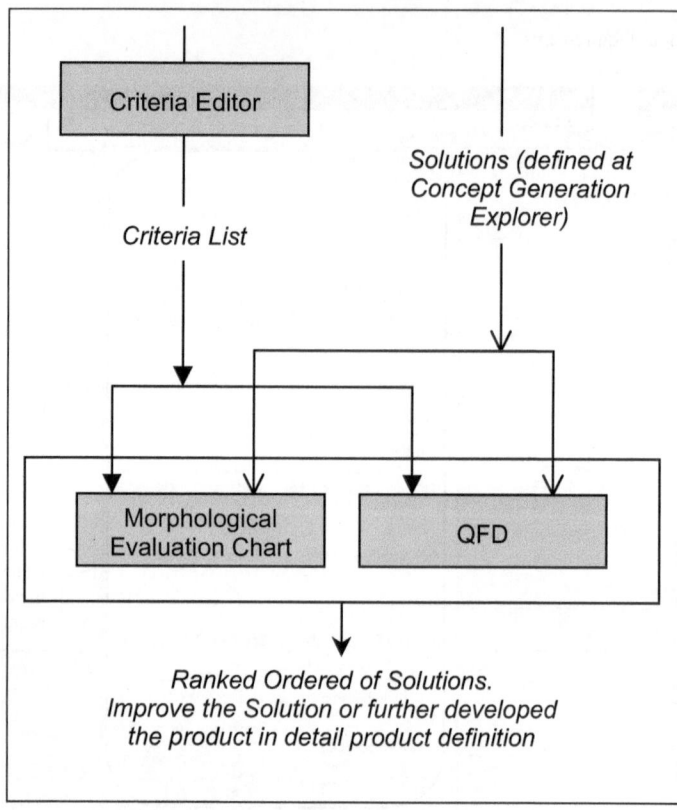

Figure 12.9 Overview of Concept Evaluation Explorer.

12.5.1. Establishing Evaluation Criteria

Criteria play an important role in concept evaluation. Both morphological evaluation chart and QFD need a list of criteria in the first column of the matrix. The Criteria Editor shown in Figure 12.10 is provided for establishing a criterion list. The Criteria Editor is easy straightforward to use. The Criteria Editor is divided into two sides. The left-hand side shows the structure of the criteria list. The right-hand side displays the details of the criteria. Analysts can modify the names or weighting of the criteria on this side. However, there are a few issues to be resolved.

The first issue is how criteria are represented in both user interface and backend database. In this work, a hierarchical tree structure is adopted for grouping the evaluation criteria into categories, sub-categories, and so on.

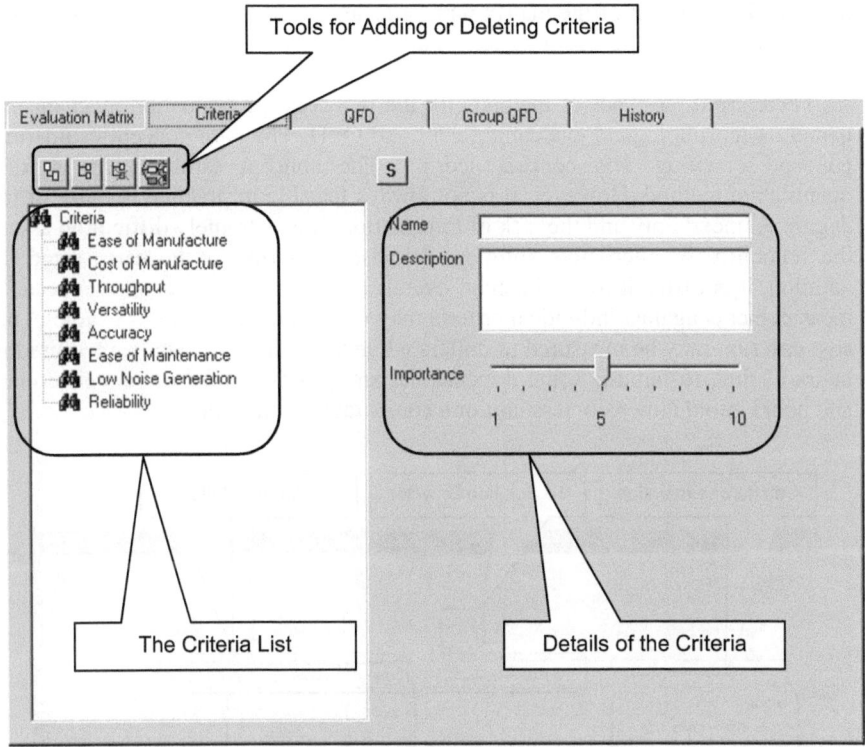

Figure 12.10 The Criteria Editor.

The second issue what should be collected in the criterion list. These criteria are chosen mainly based on two dimensions: customer needs and enterprise needs (Ulrich and Eppinger, 1995). Customer needs, such as ease of use, and ease of handling etc., are identified by analysts or through market research. Enterprise needs are requirements generated by enterprises. Some examples are low manufacturing cost, ease of assembly and minimal risk of product liability.

Thirdly, criteria are not equally important in most cases. Some criteria are more important to the future success of a product than others, and this strongly influences the choice of designs for further success of the product. Therefore, weighting of criteria is necessary to specify the importance of the criteria.

12.5.2. Comparing Alternative Solutions with Morphological Evaluation Chart

The basic mechanism for comparing alternative solutions is a morphological evaluation chart, as shown in Figure 12.11. The vertical dimension of the matrix represents a list of evaluation criteria as compiled using the Criterion Editor. The horizontal dimension of the matrix represents all the solutions that are considered feasible to achieve the overall goal. This solution list can be directly transferred from the Concept Generator through the Concept Base.

There are three types of methods for the two concept evaluation mechanisms, namely a morphological evaluation chart and QFD. They are concept estimating, concept screening, and concept scoring. The concept estimating is a fully quantitative method. However, it is not always feasible in practice because of the degree of uncertainty and the lack of information. It is extremely difficult to assess the reliability of alternative solutions in absolute terms at the early stage. In addition, it is difficult to establish an overall measurement for a solution because measurements against individual criteria may use different units. For example, the cost criterion may be measured in dollars while the reliability may be measured in hours of time-to-failure. Some dimensional analysis is needed to convert dollars and hours into a dimensionless unit or a comparable dimension.

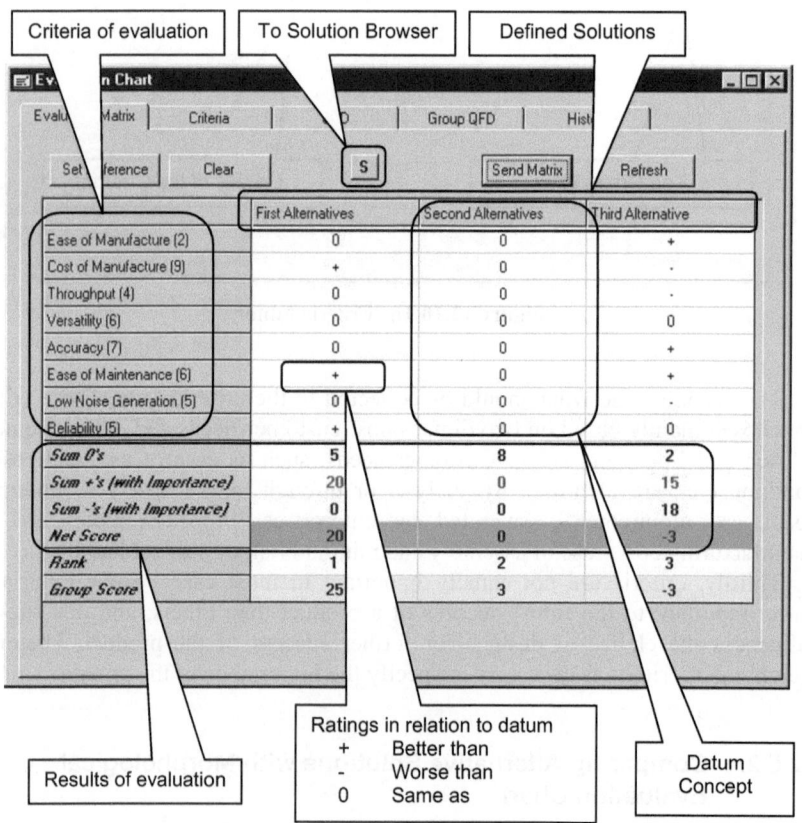

Figure 12.11 The Morphological Evaluation Matrix.

In order to overcome the above- mentioned difficulties, simplification is necessary for the evaluation mechanisms. Concept screening is one simplification. The structure of the concept screening and concept scoring is exactly the same as the concept estimating. The only difference lies in the method of measurement. The first two are measured qualitatively while the latter one is measured quantitatively. The qualitative method of measurement is to compare concepts with qualitative terms, such as 'better', 'worse'. 'same', 'related' or 'not related'.

A compromise between quantitative and qualitative method is the method of concept scoring which is a semi-quantitative method for evaluating concepts. Scores and ratings of various scales can be used. The Morphological Evaluation Chart and QFD used in the Concept Evaluation Explorer use the concept scoring method to help design teams to evaluate the concepts.

This method is particularly effective for narrowing down the number of solution alternatives, so as to improve them. If a satisfactory solution can be sought, then further in-depth investigation will be conducted to it. If no solution is found satisfactory, then the project will backtrack to the previous stages to generate more solution alternatives. If a potential solution is identified, it will be revised or combined with other solutions. If necessary, it will be re-evaluated for acceptance.

Figure 12.11 shows a sample Web page of the concept screening matrix. A datum solution is selected as the reference for evaluation against other alternatives. Names of the defined solutions and the evaluation criteria are listed horizontally and vertically on the chart respectively. The weightings of the criteria are shown in the parenthesis. If a solution is better than the datum in terms of a specific criterion, then a plus sign "+" is entered into the corresponding cell in the matrix. If a solution is worse than the datum, then a minus sign "-" is entered into the corresponding cell in the matrix. If a solution is equally good or bad as the datum, then an equal sign "=" is entered into the corresponding cell in the matrix. The total numbers of "+", "-" and "=" are counted for each alternative solution respectively. Solutions with the most "+" are deemed for further investigation.

12.5.3. Evaluating Individual Design Features Based on QFD

While the Morphological Evaluation Chart compares the solutions against a set of the defined criteria, QFD in the Concept Evaluation Explorer is responsible for evaluating the engineering details of individual each solution individually. Figure 12.12 presents an overview of the Concept Evaluation Explorer based on QFD. Across the top axis are all selected means of the solution to be evaluated. These details can be taken directly from the solution explorer. On The vertical axis of represented in the first column contains is a list of criteria against which evaluations are carried out. The weightings of the criteria are also listed vertically on the vertical axis in the parenthesis. This list can be either the same as that used in the Morphological Evaluation Chart or established separately.

In the chart, different ratings: 1, 2, and 5 have been used to show whether a particular means has strong, medium, or weak influence on a design feature respectively.

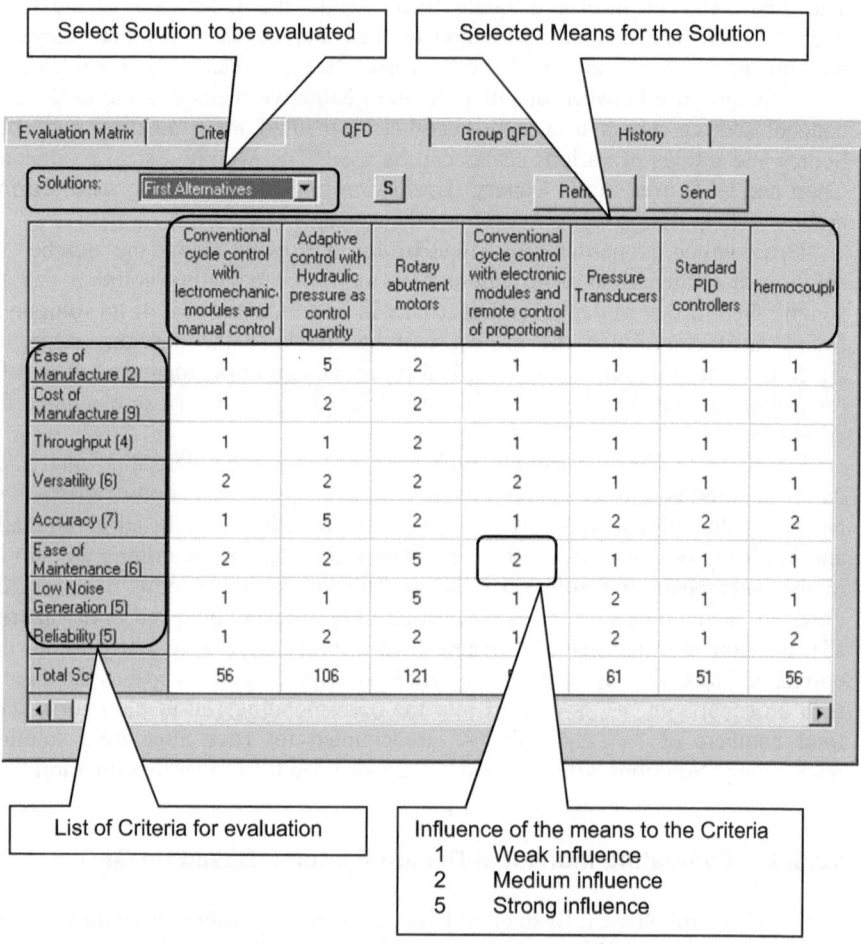

Figure 12.12 QFD on the Concept Evaluation Explorer.

QFD can be carried out independently from the morphological evaluation. However, they can actually be related to each other. The sums of the QFD rows against the evaluation criteria can be directly entered into the corresponding column in the morphological evaluation chart. The overall total in the QFD matrix for a solution should be the same as the overall total of the corresponding column.

12.6.   SUMMARY

This chapter has reported on the development of a prototype web based system for the early product definition, and of course its underlying methodology. Here is a summary of the key contributions, potential applications, and directions for further investigation.

12.6.1.   Key Contributions

As compared with these related works, our work contributes in a number of distinctive ways. First, we have not only further structured the methods involved in concept design, but also have computerised them in an easy-to-use fashion. Without computerization implementation, paper-based versions of these methods are extremely time-consuming to use. For example, redrawing may be necessary if a new function is to be inserted into the hierarchical function tree. But with computerised systems, new items can be added and obsolete items can be removed at any time and the system can automatically reconstruct the various charts used for concept design.

The second contribution, which is related to the first one, is that the computerised modules are put into an integrated framework so that they can be triggered from one to another. Forward and backward tracking are made simple and easy.

Third, the morphological generation matrix is an intuitive version of the mathematical mapping process used in Suh's axiomatic approach. Such mathematical algorithms can be built into the morphological charts to improve their level of automation if necessary.

Fourth, the concept base is built to capture conceivable concepts relevant to a specific domain. This is made possible only with a computerised system.

The last, perhaps the most significant, contribution is the use of Web-based Internet/Intranet technology to computerise the system. The system available on the Internet/Intranet appears exactly like the stand-alone version with the same functionality. Installation and maintenance are no longer necessary on the client side. As long as the user has the use of a Web browser, he or she can have instant access to any Web-based design tool available on the Internet/Intranet. The client-server architecture is a metaphor of collaborative product development, especially when tools or teams are geographically distributed. Both the client and the server can communicate to each other using a standard HTTP (HyperText Transfer Protocol), regardless of their hardware configurations and operating systems.

12.6.2.   Potential Applications

The method of morphological chart analysis has been itself applied on in concept design of not only products, but also production processes and manufacturing systems of a concept design. Many case studies have studied existed on the use of morphological chart analysis for product design (Hubka et al, 1988). However, the method is not as popularly applied in process planning and manufacturing system design as for product design, despite its suitability. Nevertheless, increasing attention of the method is drawn to conceptual design of manufacturing processes and systems. The work done by Rampersad (1994) is one such an example, using the method for assembly system design.

The use of the method for manufacturing system design is similar to that for product design, simply considering the system as a special type of product consisting of various components. When it is used for process planning, the manufacturing features of the product become the "goals" in functional analysis and the company's manufacturing processes and business activities become the "means" in the morphological chart analysis. Both goals and means are related to each other by manufacturing capability data in the concept base. A number of overall manufacturing process plans can be generated using the combination charts and the evaluation charts can be used to identify the most appropriate candidate.

Moreover, the Web-based morphological chart analysis has incredible effect on team-building. One the one hand, morphological chart analysis, as one of the formal design tools, cultivates and nurtures teamwork and collaboration. On the other hand, teamwork and close collaboration are important elements for successful implementation of this design tool. The Web-based implementation contributes to both. With the help of Web-based morphological charts, collaborative efforts can be made for generation of alternative solutions to a design problem and selection of the most appropriate solution. The system can used to solicit internal contributions as well as external contributions. More importantly, these Web-based techniques facilitate participation and involvement of customers and suppliers in the development process. For example, the Functional Analyser can be properly extended for acquiring and analysing customer requirements directly from customers. In addition, potential suppliers and experienced customers can use the concept browser and editor to express their views so as to contribute to a specific project.

12.6.3. Limitations and Future Research

Have listed main Despite of key contributions and potential applications, the present prototype system suffers from a number of limitations. First, the Concept Base at present is limited in its coverage. Its contents are slim It contains only half dozen of assembly functions and about forty means to achieve these functions. For a system is to be of practical use, comprehensive extensions are necessary to include more assembly functions and potential means. The Concept Base is also rudimentary in its database structure. Goals, means, and their relationships are described in their simplest forms. Although there are sufficient data for demonstration and simple applications, more details must be captured to ensure in the concept database so that the resulting system can works in a more intelligent way. The second limitation, also related to the concept base, is that all information, including goals, evaluation criteria, intermediate and final solutions, and results are all must also be stored in the Concept Base. a database. At present, the concept base is used for this purpose. However, setting up a separate database may be more appropriate. The third limitation lies in the way of in which individual modules' work, and which has been discussed in relevant sections. Major issues include establishing the criteria for retrieving conceivable means, short-listing feasible means, and formulating overall solutions. Finally, the system only works on the Intranet because of the limitation of its programming environment, as data binding facilities do not support Internet-based applications at present.

There are several possible directions for the improvement of the prototype system and overcoming the above limitations. Our future work would focus on overcoming the first and third limitations that are related to each other to some extent. The structure of the Concept Base would be extended to incorporate more property tables to describe details of individual goals, means, and their relationships. The criteria for retrieving conceivable means, short-listing feasible means, and formulating overall solutions would then be investigated to reflect such extensions. It is envisaged that some of such extensions would involve new techniques and concepts emerging from the research into knowledge- based expert systems and artificial intelligence. A substantial amount of further work would be involved in implementing these extensions.

Further work is necessary to overcome the final limitation that the prototype system is only available on Intranet only. However, no great difficulty is foreseen for achieving this because the problem would be solved as if another programming environment with such Internet capability is chosen for implementing the system.

13

WEB-BASED COLLABORATIVE PRODUCT DESIGN REVIEW

Design review is a vital control point for a design project to transit from one stage to another. It involves gathering and evaluating details of a product design and the concrete plans for realization and it, suggesting improvements, and confirming that the process is ready to proceed to the next phase. Design ideas that are generated from one design stage are first submitted to the design review system. If they are confirmed to be good design ideas, then they will be are released to the next stage for further development. However, if shortcomings are identified in with these ideas, then they will be are revised and refined. This review-revise-release cycle is shown in Figure 13.1.

Product design review is one of the typical scenarios of collaborative product development. A team is tasked with the design and development of a new product. It consists of members who are typically geographically dispersed. Some are lead users (key customers), some are core (key) suppliers, and others may come from various functions and units of the organisation.

Traditionally, design review is conducted by circulating documents of the product design, so that they can be reviewed one by one. After that, a meeting is then arranged to resolve different opinions. This process is very inefficient, especially when some external members from other regions such as key customers and suppliers are involved.

Despite the fact that Information Technology (IT) has been employed to facilitate this process, it is still far from satisfactory. This observation has been verified by our contacts with industrialists involved in both IT and manufacturing sectors. The roles of word processors and email exchange are limited. Clearly, a new paradigm is needed to completely reengineer this vital process of design review.

The aim of this research is to develop a methodology so as to build a more efficient and effective design review system and to demonstrate the framework through a prototype web-based platform on the Internet/intranets.

This chapter reports on the progress of resolving issues related to the development, implementation, and application of such a web-based design review framework. Section 13.1 provides a brief review of the related work in the literature. Section 13.2 will present some development considerations of the framework. Section 13.3 will use a simple example to demonstrate the application of such framework. Section 13.4 will briefly highlight some of the implementation issues. Insights and lessons will be drawn to conclude the chapter.

13.1. LITERATURE REVIEW

The literature on design review is very limited. Comprehensive searches have been carried out on the Internet and library collections. Only a few items have been found relevant. It has to be acknowledged that most of the textbooks on product design and manufacturing do mention this topic of design review as a vital activity in the product development and realization process. But the extent of coverage is usually very brief.

Over the years, different authors, practitioners, and organizations have presented various phases of design review although design review takes place whenever it is necessary and useful. For example, Dhillon (1996) suggests three major review phases:

- Preliminary design review. The main focus at this stage is on the review of the product design specifications established after the market research. This is usually considered early design review, often including the review of the conceptual product design.
- Intermediate design review. This review takes place between the two design stages of the conceptual product design and detailed product. Preliminary layout product design is normally available for review at this stage, together with the key component designs and key characteristics of key components.
- Final design review. This is often called the critical design review because it is usually performed soon after the production drawings (and design decisions) are completed for all the components and parts. If the product design passes this review, it is released for production.

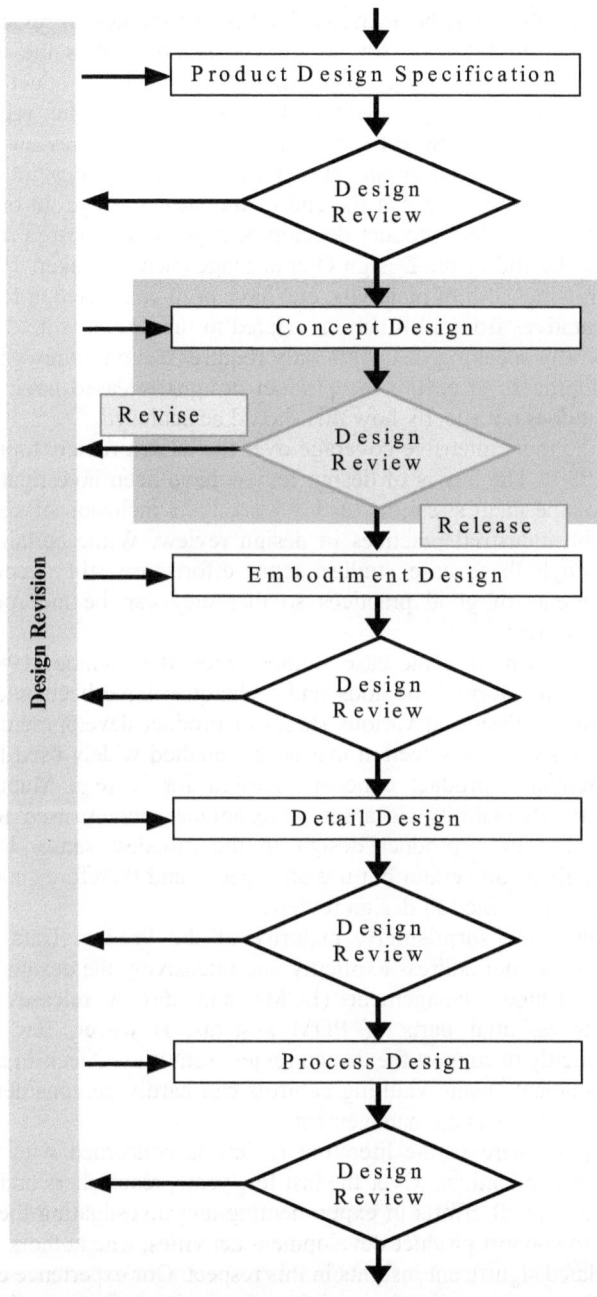

Figure 13.1 Review-revise-release cycle in product design process.

Another fact that must be acknowledged is that the design review is assumed to be mandatory in ISO9001 for design verification across the entire product development process. The design verification for a specific product design should concentrate on determining whether the design meets the requirements or specifications. Although ISO9001 indicates that the design review meetings can be held at different, appropriate stages of design, the design review being presented is assumed to occur at the end of the Design Stage. In other words, as part of the ISO9001 ideal product development process, a design review activity takes place at the end of the Design Output Stage (Schoonmaker, 1996). Once all the design drawings, bill of materials, etc. have been completed, a review meeting with representatives from all functions related to the design is held to review the design. Generally speaking, ISO9001 only requires design reviews in the product development process to ensure the product designs released having satisfactory quality. But it does not specify how this should be achieved.

Perhaps the most intensive coverage over the design review topic is the work by Ichida (1989). The basics of design review have been investigated in relative depth. One of the main strengths of this work is its inclusion of six case studies concerning the industrial practices in design review. While certain insights are provided through these case studies, extra efforts are still needed to extract common elements of good practices so that they can be incorporated into a computerised system.

It is well known from the case studies reported in Ichida (1989) and other survey reports that formal methods and techniques have been widely used for reviewing product design at various stages of product development process. For example, Pugh's concept selection matrix is a method widely used for evaluating proposed alternative product concepts. Design for X (e.g. Manufacturability, Assemblability, Realiability) is a set of techniques widely used for evaluating manufacturability of a product design in the broadest sense. Usually, these methods only focus on certain but not all aspects, and therefore should hardly be considered to replace the full design review.

Interestingly and surprisingly, majority of the Product Data Management (PDM) systems do not address explicitly and intensively the design review issue. Engineering change management (ECM) and design releases are usually considered as essential parts of PDM systems. However, few facilities are provided explicitly to support the design review activities. Check-in and check-out of design documents using vaulting controls can hardly be considered as design release, review and revision management.

Another point here in the literature review is concerned with the technical aspects of web applications. Over the last few years, researchers and practitioners have spent substantial efforts in experimenting and investigating the uses of web and Internet to support product development activities. The authors' research has also accumulated significant insights in this respect. Our experience encourages us to explore the challenge in the web-based transition from product design to manufacture through online design review.

Finally, it must be acknowledged that forward-looking companies have applied web sites to support design reviews. For example, Pfund (2001) reports that a web site is set up for each design review and members of the review committee assigned by the project manager are able to access the design documents at the web site. This type of design review practice is expected to prevail in the near future. The central subject matter of the research reported in this paper is concerned with how such approach should be applied in a systematic manner to maximize the benefits and to minimize or even eliminate some limitations associated with the traditional DR practices.

13.2. FRAMEWORK FOR COLLABORATIVE PRODUCT DESIGN REVIEW

Design review is used throughout the product development process to evaluate the design in terms of costs, quality and delivery, to ensure that the most suitable knowledge and technology are incorporated into the design, and to resolve possible problems instead of passing them downstream. A rigorous and systematic approach is needed to achieve this purpose. The systematic design review has three aspects:

- Systematic design review procedure. Different companies follow different procedures of design review. Whatever forms they take, they have a basic review pattern (Ichida, 1989; Voigt, 1996): (1) Collecting and compiling information; (2) Defining evaluation criteria or targets; (3) Evaluating product and process designs and supporting operations; (4) Proposing improvements; (5) Defining implementation actions; (6) Confirming readiness for the next stage.
- Comprehensive design review documentation. The documentation system for design review is of pivotal importance to its effectiveness and efficiency. If excessive amount of paperwork or documentation is involved, the efficiency of design review may be adversely affected although the effectiveness may be ensured. Conversely, if insufficient paperwork or documentation is involved, the effectiveness may be compromised, while communication may be adversely affected, thus, the efficiency is ultimately affected.
- Dynamic organizational structure. Design review is always team based. Different disciplines are usually represented in the design review committee including the design and development department or unit, Manufacturing and Industrial Engineering, Inspection and Quality Control, etc. How these personnel are organized and structured depends on individual companies. Some may establish special-purpose design review department formally while others call upon a group of people in an ad hoc fashion as necessary. Some may have full-time design review coordinator(s) while others appoint temporary staff to look after specific projects when necessary.

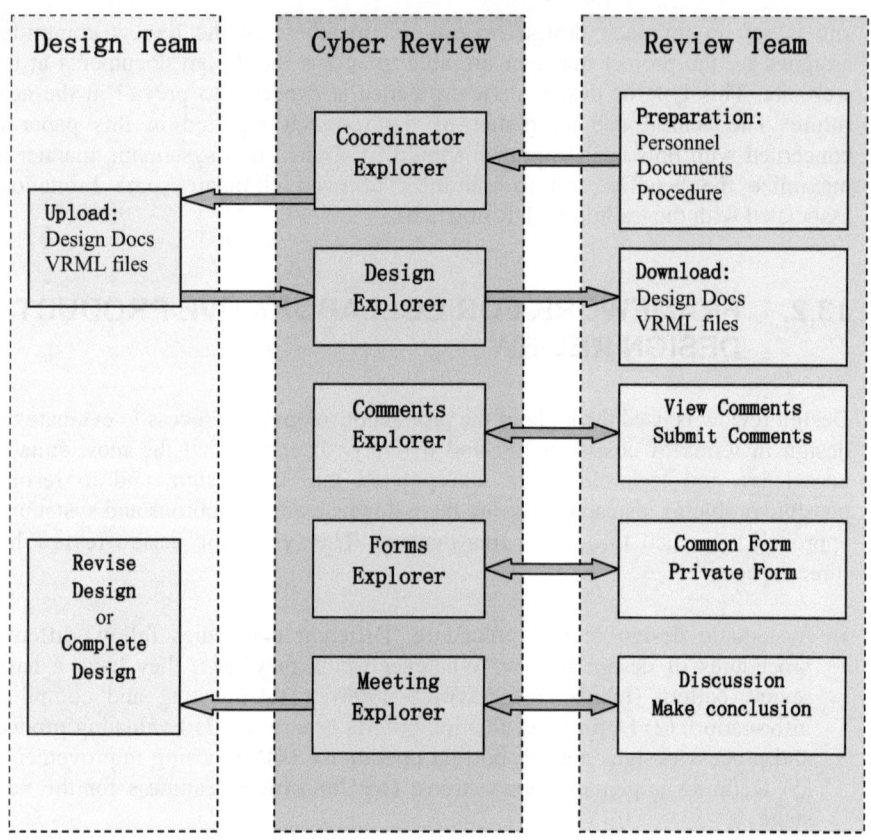

Figure 13.2 CyberReview framework.

13.2.1.   Design Considerations

Numerous factors must be addressed during the implementation of the CyberReview facilities. Some of them are:

- The industrial practices in design review vary widely from one company to another. The web-based collaborative product design review framework must have sufficient flexibility to incorporate and reflect good practices across the industrial spectrum.
- Design review naturally involves 3D geometries of the product features in addition to other textual descriptions. It has been decided to employ the VRML format in the CyberReview for this purpose.
- Comments made during the design review process are usually specific to the perspectives or positions of the 3D product drawings.

- Design review is usually interactive and collaborative. Discussions and arguments are usually conducted based on topics or threads. The electronic forum is adopted in the CyberReview for this purpose.
- Design review can be either asynchronized or synchronized or both. Synchronization poises a great challenge in implementation.

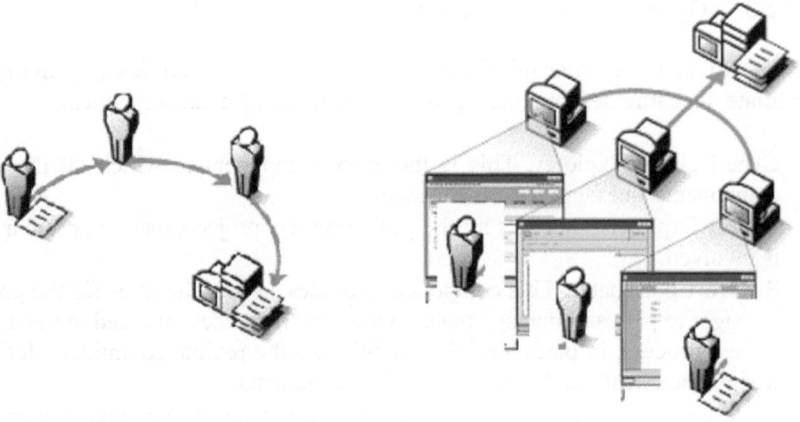

(a) Sequential design review.   (b) Parallel design review.

Figure 13.3 Sequential versus parallel review.

13.2.2. CyberReview Framework

After initial investigation, a prototype system called CyberReview has been developed. Figure 13.2 shows an overview of the CyberReview system. CyberReview is a web site that serves two main groups of users. The first group is designers or product development teams who submit designs (in the form of documents) for review. The other group is the committee established for reviewing design projects. CyberReview provides a repository for archiving the design and review documents for both groups. The memberships of these two user groups are beyond the scope of this research.

CyberReview enables the practice of design review to be reengineered from what is traditionally sequential to parallel. In the sequential design review practice, one member submits a package of design documents, as shown in Figure 13.3(a). This package is then circulated among the review committee members one by one. After that, a review meeting will then be organized. This process is usually very tedious and the time of a review cycle is very long.

In the parallel review practice, as shown in Figure 13.3(b), once a member submits a package of design documents onto the CyberReview repository, all other members of the committee are able to download and review them. All the review results will then be submitted onto the CyberReview repository. If the system detects the deadline, by which any member has not yet submitted his or her review, a reminder message will then be sent to him/her. If the system detects that all the members have completed reviews, then a review meeting will be organized either virtually or phyically.As a result, the review cycle time is reduced dramatically.

13.2.3.   CyberReview Components

As shown in the middle of Figure 13.2, a rich set of facilities is provided to facilitate the entire design review process which has 10 main components:

- CyberReview Explorer. This is the entry component to which all the other components and facilities are attached.
- Project Explorer. This lists all the design review projects that a particular user is involved.
- Review Coordinator. This component provides a set of facilities for the project manager to plan and manage the activities and resources involved in the design review process, in particular, for establishing the review committee, defining design documents, and preparing review documents.
- Design Explorer. This component is provided for the product development and design team to upload the product design onto the CyberReview database for future access. The review team uses the similar facilities to download design documents for review.
- Comments Explorer. This is an electronic discussion forum for review team members to submit their comments freely concerning the design documents submitted by the design team.
- Review Explorer. This collects all the online and offline forms and documents that are prepared during the review process.
- Meeting Explorer. This is a sophisticated module providing a variety of facilities to support holding review meetings for both the chairperson (project manager) and the team members before, during and after the meeting.
- BOM Explorer. This is a module associated with the design explorer. BOM is treated as a special type of design document required for design review when it becomes available. VRML files, comments and reviews may be directly related to BOM items.
- VRML Whiteboard. This is a whiteboard based on the VRML display of product features. This module would be necessary during the review process.
- CyberReview Utilities. In addition to the above main modules mentioned above, CyberReview provides a set of general utility facilities such as user registration, search engine for necessary information, etc.

13.3.   CASE STUDY

This section will use an example to illustrate the procedure and facilities of the CyberReview system. The sample product is a push scooter which has been used extensively by the first author for teaching product design and manufacturing. Assume that a design team (a group of students) has just completed a draft design and submitted it for review. Another team will be organized to review the design.

The review procedure using the CyberReview system generally follows that of Figure 13.2. The following main activities are included:

- With the help of the Review Coordinator, the project manager establishes a review team or committee and prepares review documents (pro-forma and procedure).
- The design team uses the Design Explorer to upload the design onto the database in the VRML format.
- Individual members in the review committee use the Comments Explorer to carry out their reviews by submitting comments and suggestions to the CyberReview database. This process is generally asynchronised.
- With the help of the Meeting Explorer, a review meeting is called upon to resolve the comments from individual members.

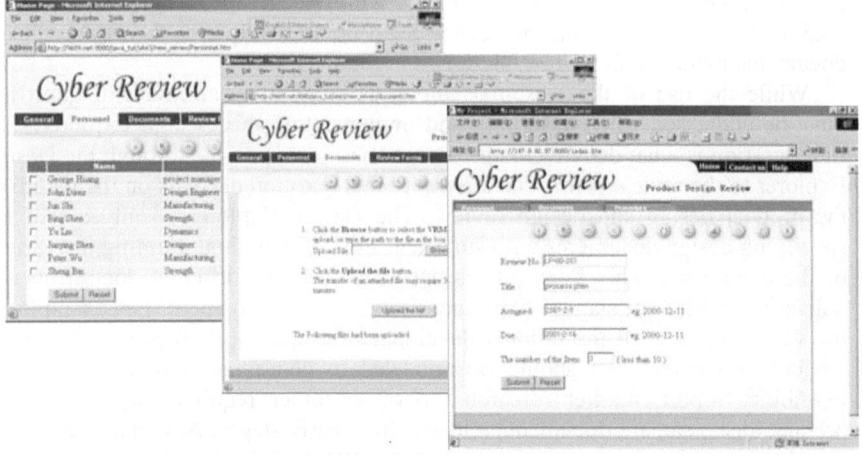

Figure 13.4 Review Coordinator for organizing the review project in terms of documents, personnel, and procedure.

13.3.1.   Organize Design Review with *Review Coordinator*

Reviewing a product design is a project (or a work package) within the product development project so, the project management techniques are applicable. The Review Coordinator is a module provided in the CyberReview system for managing the Design Review project as shown in Figure 13.4. It concerns with the following aspects:

- Design Documents. The manager must define design documents that are needed for the review.
- Review Viewpoints. The manager must define the viewpoints from which the design documents are reviewed.
- Review Team. The personnel who should be involved in the review and their individual responsibilities must be clearly defined and assigned by the manager.
- Review Documents. Review forms can be selected and retrieved from the template library. Alternatively, they can be defined according to the specific requirements of the review project.
- Review Procedure. The manager should define the procedure or the workflow that the review team and individual members should follow to conduct the review.
- Upload/Download Documents with *Design Explorer*

The design team must provide all design documents for the review team. This is achieved through the Design Explorer that serves both the design team and the review team, as shown in Figure 13.5. On the one hand, the design team uses the Design Explorer to upload the documents with the facilities as shown in Figure 13.5(a). On the other hand, the review team uses the explorer to browse all the documents included for review, as shown in Figure 13.5(b).

While the use of the Design Explorer is straightforward, several questions must be addressed when designing and implementing this Design Explorer. The first question is what documents the design team must provide through the Design Explorer to the review team. The answer to this question depends on the nature of the product design and design review. The Design Explorer is only a tool for specifying design documents with sufficient flexibility without limiting the variety of the documents. This flexibility is necessary because different documents are delivered at different stages of product design and development. For example, at the very beginning of the product development project, a draft proposal may be the only document that should be reviewed to determine if it is worthwhile. Feasibility report, market research report, customer requirements, conceptual product ideas, etc. are documents released from early stages. At certain stage, it is possible to produce what is usually called a BOM (Bill of materials). This is a piece of vital information. Similarly, design drawings are needed for review once they become available.

The second question is concerned with which electronic formats the design review documents should follow. This is actually a very complicated issue. Basically speaking, the original formats should be used. However, this imposes some limitations. For example, the user often has to use the original system to read and view the content in the original format. An alternative is to convert the document into certain standard formats so that the user can access them without any particular technologies. For example, PDF (Portable Document Format) is widely used to represent documents. In product design drawings, there have appeared a number of graphical standards. Intensive research has still been conducted to generate more general standard such as STEP (STandard for the Exchange of Product model data) (http:/www.nist.gov/sc4; Hardwick and Spooner, 1998).

In CyberReview, it was decided to use the VRML format to represent design ideas. There are several reasons for this choice. Firstly, VRML files can be directly produced from most leading CAD systems. Secondly, leading web browsers support VRML without any limitations. Users can carry out various operations with the VRML files. Thirdly, the upload and download of VRML files are straightforward.

13.3.2. Submit Reviews with *Comments Explorer*

CyberReview provides two different ways of submitting comments and reviews:

- Comments and reviews as specified by the review documents and forms in terms of the contents and formats. This will be discussed in the next section.
- Comments and reviews as interactive online discussions. This will be discussed in detail below in this section.

The Comments Explorer, shown in Figure 13.6, is basically implemented as an online electronic forum or group discussion. Discussion threads usually start with review documents submitted by the design team as required. For example, the reviewer may choose to start commenting on the product BOM by clicking on the BOM hyperlink in the Design Explorer. A list of comments is then displayed in another browser window. The reviewer is able to take several actions at this point:

- Look at the detail of an exist comment. This is a straightforward action by just clicking on the title of the comment. The content of the comment will display in a new web page.
- Create a new topic/comment. A new topic or thread is the top-level comment made to a design document submitted for review. One design document may have multiple comment threads. The creation of a new thread is the same as the response to an existing comment except that all the properties and details of the thread are specified. They include the title of the comment, type of the comment, VRML file and the viewpoint if necessary, and the content detail.
- Respond to an existing topic or comment. A reviewer is able to respond to an existing topic/thread or comment by clicking on the appropriate hyperlink. The reviewer enters the detail of his or her comment. (S)he may wish to make certain changes to the properties of the comment at this point. For example, the type of the comment may be changed from "Normal discussion" to "Require Team Discussion". In addition, the viewpoint of the associated VRML file may be adjusted and recorded to reflect his or her changed focus.

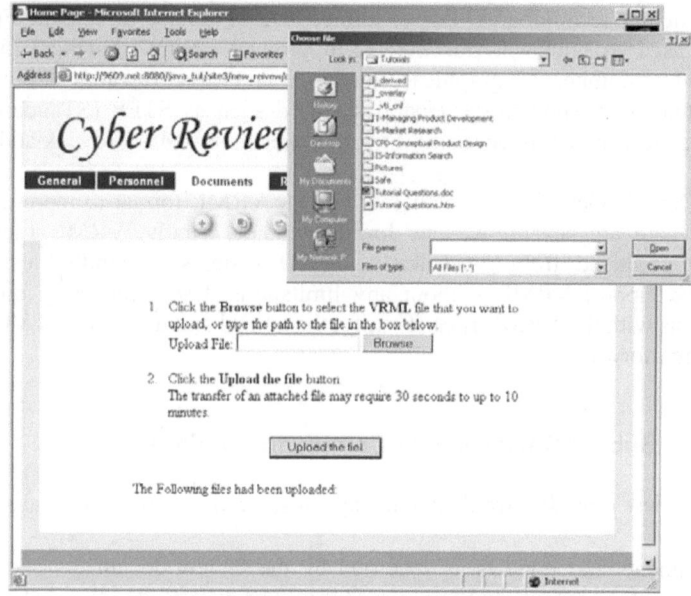

(a) Upload design documents

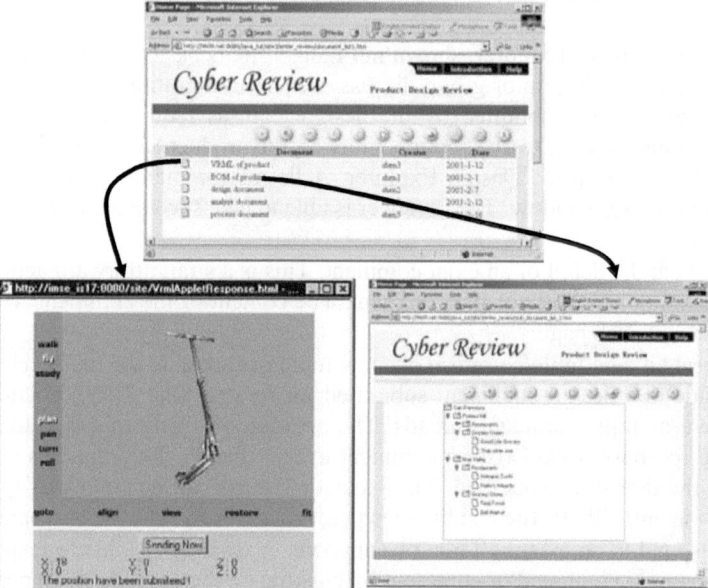

(b) Download and preview design documents

Figure 13.5 Design Explorer for uploading/downloading design documents.

Figure 13.6 Comments Explorer for submitting reviews.

Two points are worth further explanation. The first point is about the type of comment. At present, the following list is used in CyberReview although a more meaningful list must be established based on industrial good practice in the future:

- Normal Comment. This comment is just a note on a particular issue and no further action is needed.
- Require Further Cautions. This comment requires further attention/action from a specified member in the review team.
- Require Team Discussion. This comment requires further attention/discussion from all the members in the review team.
- Design Change/Improvement Ideas. This comment is a suggestion for a change in a design, requiring approval and implementation if approved.

- Defect Report. This comment is a note on a design defect or problem, requiring a solution without compromise.

Another point is that comments may be directed to items in the product BOM. There are two ways to associate comments with product items. The first method is to identify the product BOM item first and then launch the Comment Explorer to submit the reviews. This is possible because the BOM document is an online dynamic web page in CyberReview. Another method is to launch the VRML Whiteboard within the BOM page at the top-level product item. The VRML Whiteboard is then used to identify the product item first and then to launch the Comment Explorer to submit reviews by clicking this hyperlinked diagram. Basically, the VRML display and the BOM tree are the two different forms for representing the same product items.

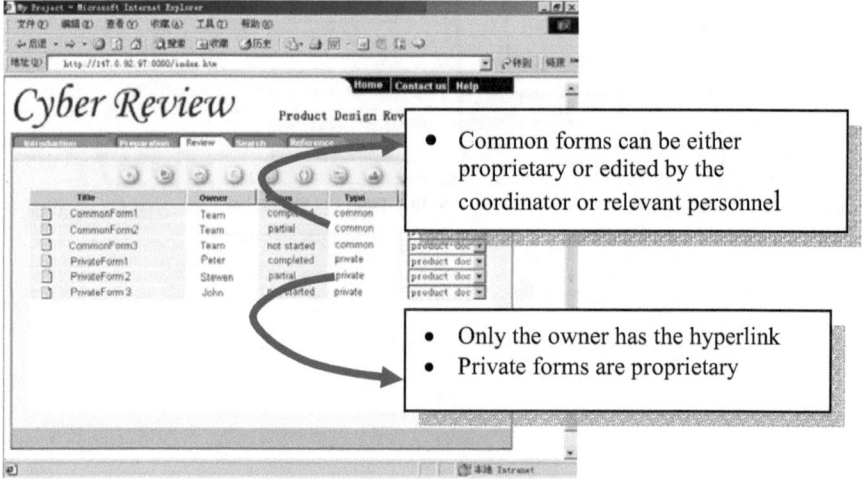

Figure 13.7 Common and private review forms.

13.3.3.   Submit Reviews with *Review Forms*

In contrast with Comments Explorer, Review Forms offer better structured ways of carrying out design review. Reviewers follow the Form Explorer, shown in Figure 13.7, as specified with the Review Coordinator at the planning stage of the design review project. There are two types of forms. At the top of the list are common forms shared by all the team members. The titles of all the common forms have hyperlinks to corresponding web pages. At the bottom of the form list are private forms owned by individual members. Only the owners have the hyperlinks to access the web pages.

Common review forms are shared by all the members in the review team to ensure a high degree of consistency between them. Common forms can be edited using the Form Editor. Alternatively, they can be proprietary web applications. A typical common form used for reviewing manuscripts for publication in journals is a checklist that all the referees must complete for the editor so that (s)he is able to make a final decision to the authors.

Private review forms or reports are usually proprietary web applications. Relevant supporting documents are attached to a review form. For example, a review based on Finite Element Method (FEM) requires a specialised form to prepare the final report. A review from the viewpoint of cost estimating needs a form different from that of FEM.

Finally, there are situations where individual members are assigned with responsibilities of completing different portions of the same review form. For example, some member(s) must provide product information in an FMEA (Failure Mode and Effect Analysis). Other members provide information on failure modes, cause and effects incrementally. This could become very complicated because allocation of responsibilities is not straightforward and it is unclear if all the members should work in a synchronous or asynchronous fashion.

13.3.4. Participate in Review with *Meeting Explorer*

Holding meetings is still a common method of carrying out design reviews. Controversial issues are discussed and resolved. Consensus decisions are finalized. This mode of design review still plays important roles even in the Internet environment although the forms may be slightly different. For example, meetings can be held virtually through teleconferencing facilities or online synchronized discussions.

CyberReview advocates two approaches to holding design review meetings:

- Provide facilities to facilitate the arrangement and conducting of the review meetings as they are held without the support of web applications.
- Provide facilities that support online synchronized group discussions.

As far as the first mode is concerned, the meetings are held in a traditional manner. However, web applications help the team members to prepare the agenda of the meeting and submit and circulate minutes of the meeting with the online facilities. The easiest way of doing this is by circulating email messages with agenda items and minutes as attachments. More advanced web facilities are dynamic web pages.

Our main focus when developing the review Meeting Explorer shown in Figure 13.8 has been on the online synchronized group discussions although teleconferencing facilities are yet to be incorporated. In fact, the second mode subsumes the first mode. That is, all the facilities available in the first mode must be made available in the second mode. However, several points are worth mentioning:

- Planning a meeting. One of the main tasks in arranging a review meeting is to establish the agenda in addition to routine tasks. A list of items is initially identified from the previous reviews submitted from the Comments Explorer. This list can then be revised and edited to reflect the most up-to-date information.
- Submitting minutes of a meeting. It is a standard practice to prepare and circulate the minutes of meeting before confirmation at the next meeting. Online facilities are straightforward for this purpose.
- Holding a meeting. It is slightly complicated to hold a meeting on the web. The chairperson plays vital roles in holding and synchronizing the review meeting. Once the chairperson declares the start of the meeting, participants are able to make their individual contributions one after another on one agenda item. All the participants can view the same contents and submit their ideas on the topics.

The roles of the chairperson at a review meeting include some listed below:

- The chairperson checks if all the participants have registered with the CyberReview Meeting Explorer, and if agenda items are completed and available for the meeting. If so, the meeting is ready to start.
- Only the chairperson has the authority to move the attention from one agenda item to the next. Hyperlinks are not available to the participants until the chairperson has released the move.
- Only the chairperson has the authority to modify the items on the agenda. Members must make their recommendations first and the chairperson decides if they are added on the agenda.
- While all participants are able to keep bidding for a opportunity to make a contribution, the chairperson pick up one participant at a time. (s)he may follow certain strategies such as random, "first come first serve", rotating, "previous speaker first", etc.
- If an item requires using the VRML Whiteboard, only the chairperson is able to change the content and viewpoint of the VRML file. Participant must submit their requests for change to the chairperson first. If a request is accepted, the VRML display will be changed accordingly. This change is shared by all the participants.

In CyberReview, a meeting is also considered as a form of review activity, just like any other activities. A unique feature, however, is that the meeting is synchronous in the sense that all the members must be present online to participate in the discussion. Like other review activities, review forms can be used at meetings for members to complete simultaneously. The distinguishing point here is that there must be a consensus from a meeting regarding an issue on the agenda.

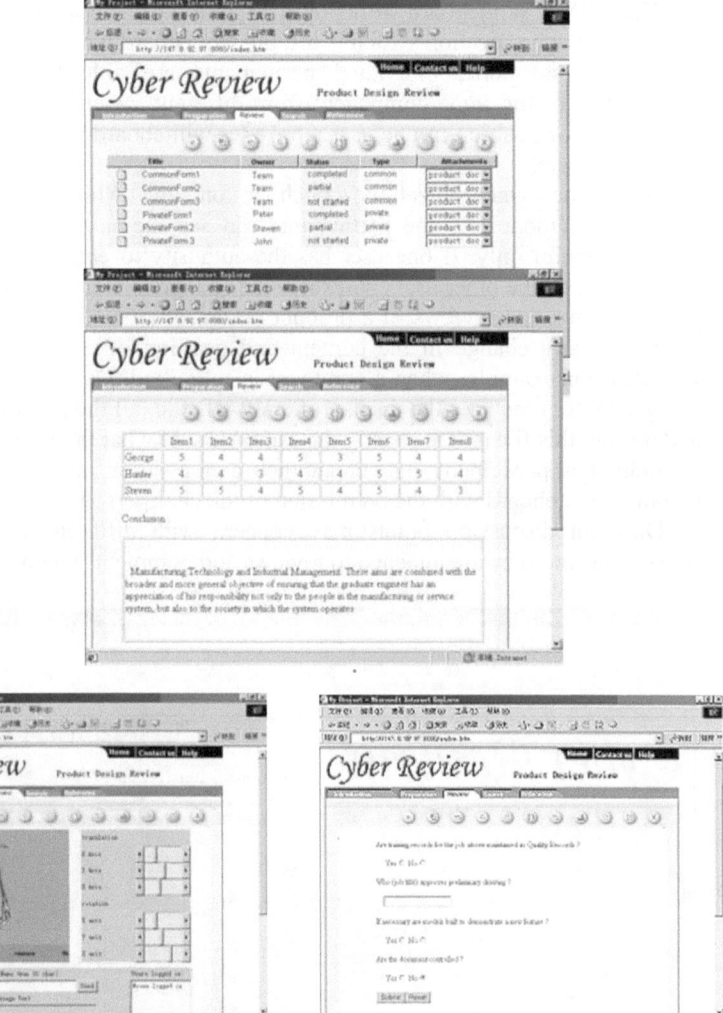

Figure 13.8 Meeting Explorer for the review team to reach a final resolution.

13.3.5.   Share 3D Product Features with *VRML Whiteboard*

The VRML Whiteboard is a main module in the CyberReview where participants can share 3D geometrical product features in the VRML format (Ames et al, 1996). The whiteboard has several main areas, as shown in Figure 13.9. One is the VRML display area as shown on the left-hand side of the screen. The right-hand side shows the position details. The third area is for defining new positions.

The VRML Whiteboard works in two modes: asynchronous and synchronous. When the synchronization switch is off, the whiteboard works in the asynchronous mode. In this mode, individual participants are able to look at different components or the same component from different viewpoints at the same time. The change of position in one user's browser whiteboard will not affect the other users' browsers.

When the synchronization switch is on, the whiteboard works in the synchronous mode. All the participants can see the same component from the same viewpoint only. If one user has the authority to change the content of the whiteboard, all the other users will share the change in their own whiteboards.

Comments and reviews of a product design often affect the design drawings. Therefore, any change in the content and position of the whiteboard must be recorded, for example, when a comment is submitted. During the meeting, when an agenda item requires the display of a VRML file, all the participants will share and display this file (share exactly the same view of what they can see).

Individual participants may create their own positions and display them on the common whiteboard with the permission of the chairperson.

Different discussion points may concern with different viewpoints of the VRML file and they are attached to reviews and comments whenever appropriate.

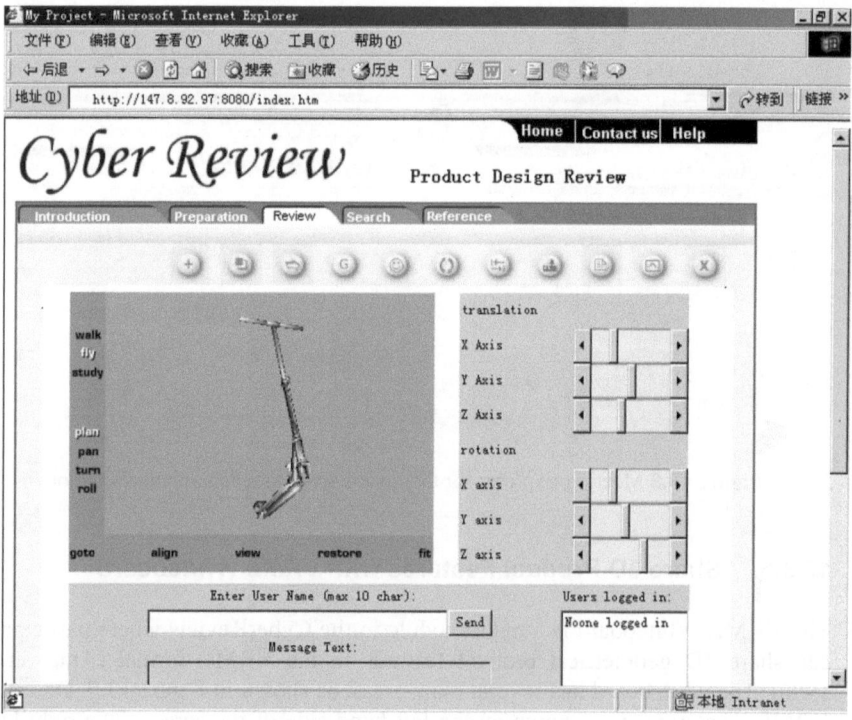

Figure 13.9 VRML Whiteboard.

13.4.   IMPLEMENTATION AND DEPLOYMENT CONSIDERATIONS

Majority of the CyberReview facilities have been implemented using various standard web and Internet constructs in relatively straightforward ways. The following considerations have been taken when implementing CyberReview:

- The standard 3-tiered architecture of web application has been used for building the CyberReview framework.
- The CyberReview application servers and database have been deployed in the MS Windows 2000 environment.
- HTML and ASP (Active Server Pages) have been used as the basic programming and scripting environment for implementing the application servers and clients, supported by Java and VB scripts when necessary.
- Java Applets and Servlets have been extensively used in the implementation of application servers and clients, especially in the synchronization of some functional facilities.
- A relational database is used. The connection to application servers is through ODBC and ADO (ActiveX Data Object). No direct connection is allowed from application clients to the database.
- The VRML (Virtual Reality Markup Language) format has been selected for displaying design concepts pictorially. Cosmo Player is used for the VRML browser although other alternatives have also been experimented.
- Electronic forum has been extensively used and modified in implementing facilities for submitting review comments, holding meetings, etc.

However, some complication has been encountered when implementing the synchronization mechanisms for the VRML Whiteboard and the Meeting Explorer. The synchronization of the Meeting Explorer has been achieved by periodically refreshing the web pages. The time can be set at 1-10 seconds.

13.4.1.   EAI for Realizing Synchronized VRML Whiteboard

The implementation of the synchronized VRML Whiteboard requires frequent communications between the VRML browser and the web browser.

Initialisation of VRML EAI by Applets

EAI (External Authoring Interface) is an API for connecting a VRML plug-in to an external program (Hartman, 1996; Roehl and Couch, 1997). EAI allows an external program to control a VRML object in VRML plug-in. In our system, the review workspace is a VRML file, and the review Applet is an external program. The use of VRML EAI requires that several classes be initialised before the connected Applets executes. Therefore the execution of collaboration Applet begins with the initialization. First of all, a browser class in initialized, it allows the Applet to access the VRML plug-in used, then the node calls in the VRML file, such as the transform and touch senser nodes. There by, while controlling the VRML object, the attribute values of the VRML nodes can be referenced or changed in the Applet. The change of a node's attribute is called an event. There are two events: EventIn and Event Out. EventIn is an event that a change attribute comes from an external program. EventOut sends out a change attribute to an external program. Both events are derived from the node class, which enables the transmission of user input on the Applet from controlling VRML object to VRML plug-in.

Sending events

Once an instance of EventIn class is obtained, an event can be sent to it. EventIn is an abstract class so an EventIn subclass is needed. This subclass contains methods for sending events of given types. If a VRML scene contains the following node:

DEF SCALER Transform { ... }

the scale can be set to values (1.0, 2.5, 1.0) from Applet like this:

```
Node scaler = browser.getNode("SCALER");
EventInSFVec3f scale =
    (EventInSFVec3f) scaler.getEventIn("set_scale");
float sc[3] = { 1.0, 2.5, 1.0 };
scale.setValue(sc);
```

Receiving events

Once an instance of EventOut is obtained, a value of the event or exposed Field can be read. Since EventOut is also an abstract class, similarly, a subclass, which contains methods for getting events of given types, is needed. Current value of the scale field can be read like this:

```
EventOutSFVec3f scale =
    (EventOutSFVec3f) scaler.getEventOut("scale_changed");
float current_sc[3] = scale.getValue();
```

13.4.2. Applet-Servlet Communication

In our application, all the user's registration information, the position information got from the VRML EvenOut and EventIn message was encapsulated into Java objects. For example, in the VRML whiteboard, when a participant changes the position of the VRML prototype, the new position message is sent to the Servlet. Also, it is our wish that the Servlet will send the Applet an updated message list as a vector of Msg objects. This will allow the Applet to display the changes of the VRML easily and the Servlet can add a choice position message to the database.

Sending object from an Applet to a Servlet

The Applet sends a position object to the Servlet when a participant changes the VRML in the COSMO player. In order to send a Java object from the Applet to a Servlet, we should make a Java object to be serializable, its class must implement the java.io.Serializable interface which allows an object to be flattened and saved as a binary file.

Sending object from a Servlet to an Applet

The Servlet can return an updated list of changed positions. The updated position list is returned as a vector of position objects. When the Servlet returns the vector of position objects, there is no need to iterate through the vector and serialize each position object individually. It can simply serialize the entire vector in one step, since the class java.util.Vector also implements the java.io.Serializable interface, in addition, the Servlet is now capable of storing the position message onto the database.

13.5.   SUMMARY

This chapter discussed a prototype web-based system for collaborative product design review in the extended enterprise environment. The system is based on a sound theoretical framework, including a systematic procedure, documentation, and organisation. Various Internet and web facilities are provided for both the product design and design review teams to carry out design review activities.

One of the major contributions of this research is that the design review framework has become completely web-based. As long as the user has the access to a web browser that is in turn connected to the Internet, he/she is able to participate in a design review project, regardless of time and place. Such a web-based implementation offers a number of advantages over paper-based and standalone computer design review systems. First, the amount of paperwork is reduced to the minimum level. Second, the throughout time is significantly reduced. This is due largely to the third advantage that a web-based design review system allows simultaneous data access and processing while paper-based and standalone systems only allow single user access. Finally, design review data are shared and communicated between all parties concerned once they enter the system.

The web-based CyberReview system described in this chapter only supports the basic design review functions and activities. There is sufficient flexibility in terms of methodology and system in the design framework to cater for wider applications in different sectors, such as mechanical products, building and construction projects, electronic components, and software packages. For example, users are able to use the system facilities to design their own design review procedures and documents, and organise the review committees. This resolves a limitation that different organisations have different design reviews at different stages of product development process.

Further developments are possible along several directions. Firstly, the design review process is part of the product development process. This relationship should be explicitly reflected in the CyberReview framework to enable natural integration. This, however, is more complicated by the fact that design reviews at different stages of the product development process have varying focuses. This merits substantial investigation in the near future.

In addition, design review process is closely related to engineering change management process. From both literature and industrial practice, it is unclear about how they are related: whether one includes the other or they exist in parallel. Further investigation is needed to see if it is possible to incorporate the engineering change management process within design review process to simplify the situation without compromising the effectiveness and efficiency.

Furthermore, design review involves intensive data of products and processes and requires substantial amount of knowledge and skills from the committee members. A central repository such as the Product Data Management (PDM) or Enterprise Knowledge Management (EKM) system is useful in meeting this requirement. When extension or integration is made in this direction, it should be noted that the scope of the design review system is further extended and the resulting system may become too comprehensive and thus too complicated to use.

Finally, design review often uses well-established tools or techniques. For example, organisations have used FMEA (Failure Mode and Effect Analysis), QFD (Quality Function Deployment), and some special tests in design review. Fortunately, there have been significant developments in web applications for formal methods and techniques. Their integrations are made as easy as showing their URLs.

Although CyberReview has not yet been put into practice for testing and verification, the basic functionality and performance have been demonstrated within the prototype system. No serious obstacles have been identified that would limit it for real-world industrial applications.

14

ONLINE COURSEWARE ENGINE FOR TEACHING BY EXAMPLES AND LEARNING BY DOING

During the last a few years, there have been significant developments in the online course tools over the world wide web (WWW or web). WebCT (http://www.webct.com), e-Teaching and e-Education (http://www.ecollege.com; http://www.e-education.com), and Course Info (http://www.blackboard.com) are some of the examples widely used at colleges. These tools host teaching and learning materials for both the teachers and students while provide facilities to enhance and improve the interactivity between the teachers and students. Typical online facilities provided in these tools include syllabus tool, calendar of events, email and live chat boxes, threaded discussion forums, tests and exams, announcement bulletin boards, study journals, feedback questionnaires, etc.

These course tools can be used within and/or outside the classrooms. If they are used outside the classrooms at distance, they provide a virtual classroom environment for delivering curriculum materials at any time from anywhere by anyone who is enrolled on the course (Turoff, 1994; Hiltz and Benbunan-Fich, 1997). If they are used as an enhancement in classroom teaching, they not only improve the accessibility but also interactivity within the classes. Wade et al (1999) discusses their experience of not only the use but also the development of web-based educational environments for software engineers.

TELD is yet another online courseware engine over the web. However, TELD has its unique features. Firstly, the TELD engine is also a courseware search engine with which both teachers and students are able to search for relevant curriculum materials. This is similar to the function of the NEEDS search engine where teachers register their teaching materials while others including students get hyperlinks to down these materials. Secondly, the TELD engine is a web host with which teachers and students archive and obtain curriculum materials. This function is common in most software course tools. However, TELD specifically support the "Teaching by Examples and Learning by Doing" method that unifies Case method (CM), Problem-base learning (PBL), and Project-based learning (PBL) widely practiced in business, medical, and engineering education respectively. The third unique feature of the TELD engine is its online facilities supporting group activities in collaborative and participatory learning. Groups are able to plan their exercises and projects within TELD. Groups are able to prepare agendas for holding workshops or meetings on specific issues of the project and then report on the progress in the form of minutes of meetings.

First of all, Section 14.1 will give an overview of the online TELD courseware engine. Section 14.2 will discuss some of the design issues related to the development of the TELD courseware engine. The general operation of the TELD courseware engine will be briefly shown in Section 14.3 using a "Product Engineering" course as an example. Some insights from the initial experiences of the TELD method will be drawn in the last section to conclude the chapter.

14.1. WWW.TELD.NET: ONLINE COURSEWARE ENGINE

TELD does not only unify the Case Method and the Problem/Project-Based Learning method but it also represents a web-based online courseware engine as a computer system on the Internet. After our initial efforts, the prototype TELD system has been developed. Figure 14.1 shows a general scenario where TELD is used to support different faculty and student users. The TELD courseware engine combines the following four key functions into one framework:

- Firstly, TELD represents a teaching and learning method that unifies a number of contemporary methods such as Problem-Based Learning (PBL) in medical education, Project-Based Learning (PBL) in engineering education, and Case Method (CM) in business education.
- Secondly, TELD serves as a web server for hosting teaching and learning materials especially based on the TELD method. A variety of online facilities are provided for editing and uploading course materials such as syllabus, schedules, curriculum, examples of case studies, exercises of mini-projects, and assessments.
- Thirdly, TELD is a courseware search engine where educators are able to register their course materials and search for suitable materials for a particular course. In contrast with general-purposes of other search engines, TELD is set

up for the special purpose of education. Therefore, the time and efforts spent on surfing are expected to reduce dramatically.

- Finally, TELD is an online virtual classroom for electronic delivery/distribution of electronic curriculum materials.

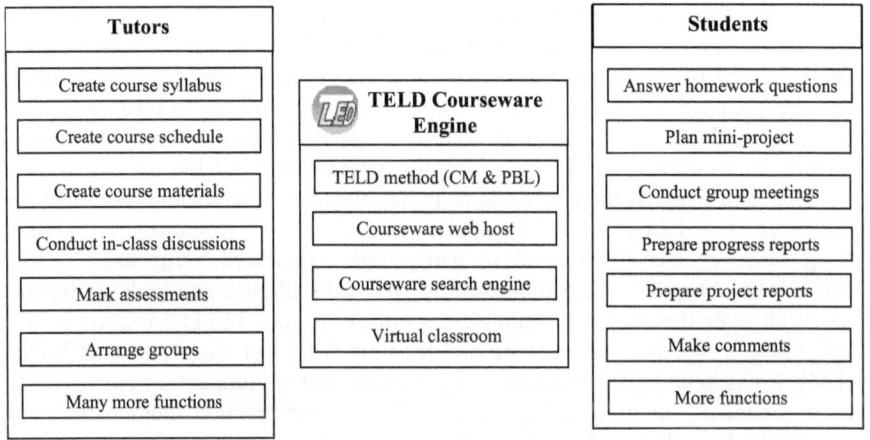

Figure 14.1 TELD overview.

14.1.1. TELD as a Teaching and Learning Method

As said in an ancient Confucius education philosophy, "I hear and I forget. I see and I remember. I do and I understand", "Teaching by Examples and Learning by Doing (TELD)" exactly reflects this ancient philosophy in modern education within a technology-intensive environment. "Teaching by Examples" allow the students to "see and then remember" while "Learning by Doing" allows students to "do and then understand".

In other words, TELD is a method that uses examples and case studies extensively for teaching a particular subject and uses exercises and projects for helping students to digest the materials. This method has been widely used by a number of leading universities in delivering a variety of subjects and courses, with a number of variations. In medical education, TELD is well known as problem-based learning (PBL). We can trace its origin back to the Faculty of Health Sciences of the McMaster University (http://www.samford.edu/pbl; http://edweb.sdsu.edu/clrit/learningtree/PBL). In late 1950s, it became a model for many other universities to follow. By the mid-1980s, many leading medical schools all over the world started using PBL. There have appeared a number of well-known PBL centers around the world, and a number of PBL conferences. The method has spread out to other disciplines such as sciences and engineering. A simple search using the "PBL" keyword on the Internet results in a rich list of relevant resources over the World Wide Web. Some of the URLs are listed in the references.

Another variation of TELD is the Case Method (CM) used in business and management education. The Harvard Business School has used the CM for almost a century as an effective way of teaching business administration. Cases are taught in a variety of disciplines around the world. A case is a description of an actual situation, commonly involving a decision, a challenge, an opportunity, a problem or an issue faced by a person or a team in an organization (Erskine, Leenders and Mauffete-Leenders, 1999). Cases are field-based. That is, a case usually comes from the real world organisation where data are collected, analysed and compiled to form the case. These cases allow the teachers and students to take on the roles and responsibilities of specific people in a specific organisation. That is, cases enable us to "learn by doing". Usually, cases are available in the form of hardcopies. Recently, the electronic form has been used increasingly for electronic delivery.   ECCH   (European   Case   Clearing   House) (http://www.ecch.cranfield.ac.uk/) and HBS (Harvard Business School) (http://www.hbsp.harvard.edu/home.html) have distributed hundreds of business cases all over the world. In engineering education, the TELD variation is often referred as Project-based teaching and learning (PBL) or simply Project Method (PM). It has been widely used in many design- related engineering curricula such as mechanical, manufacturing, industrial and construction engineering. The Autodesk Foundation has sponsored intensive research into PBL or PM (http://www.autodesk.com/foundation/pbl/research/). In addition, the method has also been widely used for teaching Design Technology courses in primary and secondary schools.

No matter which term the teachers and students use, this method achieves the teaching and learning objectives by intensively examining the subject matter of the project/case/problem, as specified in the course syllabus. Like PBL and Case Method, TELD strongly emphasizes student-centred learning by students themselves. This generally works well for senior students at year two and three or above, but it poses a big challenge for the first-year students. TELD balances this by providing clear guidance from the tutors/teachers so that the students know what they are doing and where they are getting and how to some extent. Such a balance between the knowledge acquisition and skill development is a major strength of TELD. That is, knowledge is most effectively acquired in the context in which it is discovered. "Teaching by Examples" provides such a context of problem-based learning. In addition, skills are most efficiently developed during the process by which a practical problem is solved. "Learning by Doing" creates such a problem-solving process.

14.1.2. Courseware Web Host

The World Wide Web (or web/www) has become extremely popular for tutors to provide their teaching and learning materials on the Internet. One of the most significant advantages is that students are able to access the materials on the Internet at anytime and anywhere. It is a straightforward practice to set up a Personal Computer as a web server to host the course web pages. Besides, most universities and organisations provide web hosts for individual course web pages/sites. For example, the Teaching Development Grant (TDG) from the Hong Kong University Grant Committee has supported a number of large projects for developing web sites and pages for various subjects such as civil and construction engineering, biodiversity, and industrial engineering, etc.

In addition, there are commercial courseware hosts such as WebCT (http://www.webct.com/). These systems have been specially developed to host course materials from lecture notes to assessment quizzes. Many useful tools such as syllabus tool, online chat tool, and course forum are provided.

Although HKUGC TDG projects are impressive and commercial courseware systems such as WebCT provide great convenience, they all suffer from a weakness. That is, they do not explicitly support any contemporary teaching and learning methods such as PBL, the Case Method or the TELD method.

On the other hand, cases are at present published and available from various sources depending on the subjects. HBS and ECCH are the two main centres, distributing hundreds and hundreds of business cases around the world. These cases are not readily available from the Internet even for subscription.

The above analyses highlight a gap between the current use of the web technology in education and the contemporary teaching and learning methods. The TELD courseware engine aims to fill this gap by providing a web host for TELD-based materials.

14.1.3. Courseware Search Engine

There is not doubt that tonnes of teaching and learning materials exists on the Internet.. The problem is how to find the most appropriate materials on the web. One of the solutions is to use search engines. Most search engines are for general purposes, not specifically developed for searching teaching and learning materials. Therefore, it is extremely time-consuming to surf for the most relevant materials by using keyword searches. The resulting list is usually very long and contains many irrelevant items. This often puts off tutors and students.

As a result, a special-purposed search engine for teaching and learning materials is needed. In fact, the United States National Science Foundation has established a coalition of educational and industrial partners. This coalition has developed a national database called NEEDS – the National Engineering Education Delivery System (http://www.needs.org/), to generate product development and design cases for educators, students, and practitioner engineers. The NEEDS engine has expanded into other subjects and disciplines such as chemistry. This kind of search engine is expected to play an effective role in teaching and learning.

TELD is yet another search engine with similar purpose. However, TELD is a search engine, hosted by itself, mainly devised for the teaching and learning materials although outside materials may also be included.

14.1.4.   Virtual Classroom and Study Room

TELD can be virtually considered as an online classroom or study room with teaching and learning materials provided by the TELD courseware engine. Tutors and students can archive and fetch materials, ask and answer questions, and exchange comments on relevant points, etc. Thus, TELD facilitates studies within and outside the classrooms. For this reason, TELD can be considered as a virtual study/class room. Some main TELD functionalities are summarized as follows:

- Course syllabus. TELD is an online courseware web server that hosts multiple courses. Facilities are provided for registered teachers and organizations to create new courses and define their syllabus in TELD. This leads to a community of teachers and learners who share the resources in the Course Library.
- Case management. Cases are the basic constructs of TELD. They form what is called the Case Library. As a search engine, TELD supports teachers to find most relevant teaching and learning materials for his/her course(s) and topic(s). Cases can be adopted/adapted as teaching examples or learning (assessment) exercises.
- Assessments. Teachers use TELD to prepare both formative (discussion questions as homework) and summative assessments (comprehensive exercises often in groups). Students need to submit their answers online to the TELD database, so that tutors can mark and comment on them online.
- Feedbacks and student-teacher interactions. Course forum and questionnaires are some of the common methods to obtain student feedbacks on the course delivery. Emails and online chats are examples of useful interaction and communication tools.
- Teaching and learning planning. Facilities are provided for teachers to make schedules for lectures, discussions, assignments and exercises. Similar facilities are also provided for students, especially for organizing their group work and/or meetings.

14.2.   DEVELOPMENT OF *TELD* COURSEWARE ENGINE

Based on our previous experience in developing web applications for product design and manufacture, some of the key design issues focus on the backend database design and the front-end user interface design of the web pages, in addition to the architectural issue. The following section will discuss on these three issues regarding the development of TELD engine.

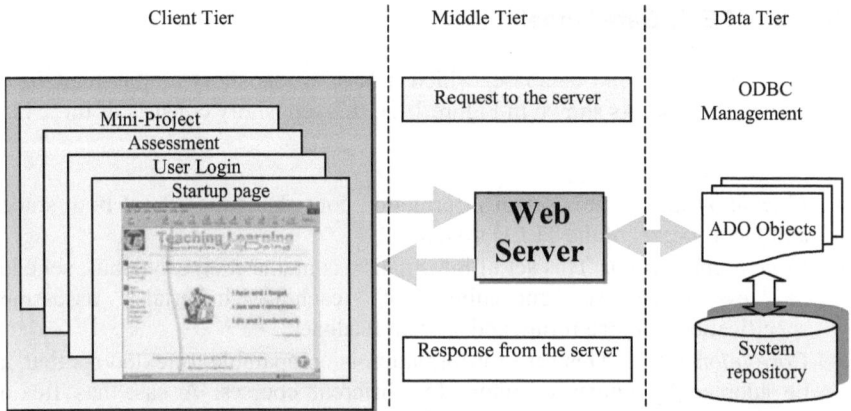

Figure 14.2 TELD architecture.

14.2.1.   Architecture of TELD Courseware Engine

System components or facilities are organised by the following typical three-tier architecture of web application, i.e., client tier, middle tier of web servers (application servers), and data tier, as shown in Figure 14.2. Application clients include client browsers with which the customers can connect to the system web Server. Web Server (Application Server) is the centre of the system. It provides the objects needed by the clients and controls the information communication between the other two tiers. The objects provide three main functions in the application: registration, search and solicitation. Data source and ODBC management provide a repository for the system. ODBC management acts as an information broker between the data source and the application, retrieves data from database by the Open Database Connectivity (ODBC) and transfers data to client by the Active Data Objects (ADO).

   TELD components have been implemented through a combination of programming techniques, such as Java Applets, Java Script, VB Script and ActiveX Objects, with ASP (Active Server Pages) as the main environment. Active Server Pages (ASP), which is considered as an enhancement of common CGI applications, is a server-side scripting environment that designers use to create dynamic web pages or build powerful web applications. It improves the capabilities, facilities and compatibility of common CGI applications. More importantly, ASP pages can call ActiveX components and Java applets to perform tasks, such as connecting to a database or performing a business calculation (http://www.learnasp.com/). With ASP, designers can add interactive contents to web pages or build the entire web applications using HTML pages as the interface to customers. For these reasons, ASP is used for implementing the WAPIP prototype. Experience shows that this is an efficient technique for achieving the desired functionality.

14.2.2.   TELD Data Model

TELD holds a backend database, which acts as a repository at the heart of the courseware engine. As shown in Figure 14.3, this repository consists of three main parts:

- *User information.* This set of information consists of records of both student and faculty users of the TELD system.
- *Course information.* This set of information contains course syllabus, schedule of lessons/sessions, curriculum of teaching materials, assessment specifications, course forum and student feedback.
- *Case information.* This set of information is comparable to textbooks that can be adopted by numerous tutors for different courses. A case has flexible number of sections, depending on the nature of the case. Some of the typical sections include Learning outcomes, Case/problem definition, Case analyses, Case summary, Readings, Discussion questions.

A course is defined by its syllabus and is scheduled by a number of lessons/sessions in a semester or equivalent. A lesson is related to a case or certain part (called "Sections" in TELD) of it. Therefore, the concept of "case" is the most fundamental building block of the TELD data constructs.

A case is usually dedicated to one key theme/topic of a course syllabus. In order to cover the entire syllabus of a course, several cases may be needed. These cases may be closely or loosely related to one another. In TELD, these cases for the same course form a "curriculum".

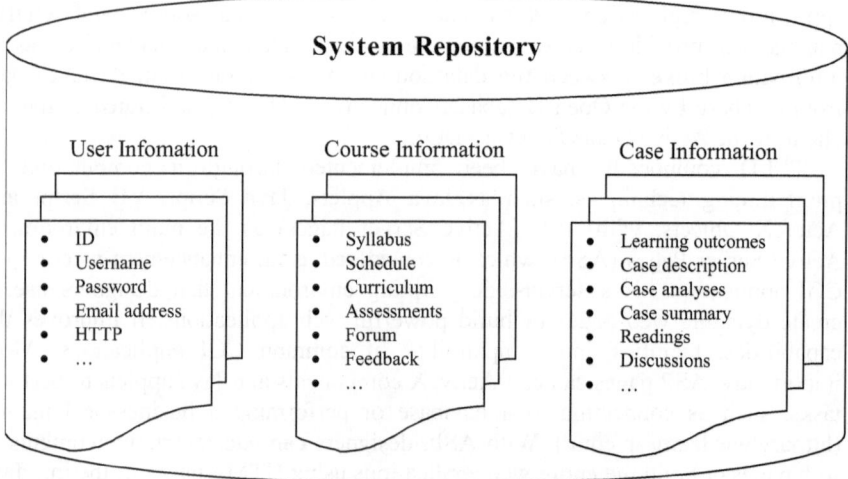

Figure 14.3 Some key data items in TELD data model.

Three basic constructs have been mentioned above for a TELD course: syllabus, curriculum and schedule. The syllabus specifies what themes/ topics of a subject should be taught/covered in a course. The curriculum specifies the content materials, such as cases in TELD, with which topics/themes are discussed. The schedule specifies when and how long the topic/theme should be discussed with the assigned content materials. Readers may have different interpretations for these terms. However, in the TELD data model, they are used with the above specific meanings.

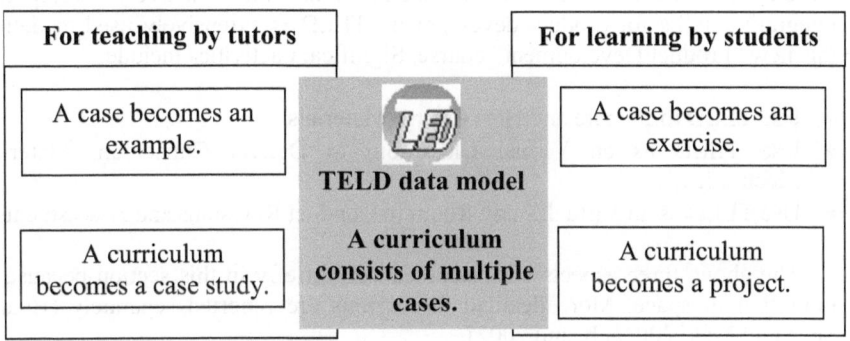

Figure 14.4 Many faces of TELD constructs.

14.2.3. Many Faces of Cases and Curricula

Cases and curricula are two of the most important data elements in the TELD data model. However, they have different names when they are used for the purpose of teaching and learning, as shown in Figure 14.4. When a case is used by a tutor to teach a course, it is called an "example". When a case is studied by students to learn a specific theme/topic of the course (as part of the summative assessment), it is called an "exercise".

Likewise, when a curriculum is used for teaching, it is called a "case study" (which consists of multiple examples). When a curriculum is used for summative assessment, it is called a "project" or "mini-project" (which consists of multiple exercises).

14.3. OPERATION OF *TELD* COURSEWARE ENGINE

It is probably the most efficient way to explain how TELD courseware engine works through an example. This section uses the "Product Engineering" and "Product Development" courses as examples. As a matter of fact, the TELD method and system is developed pre-dominantly based on our experience in delivering these courses over the last few years. Currently, TELD is used for delivering the "Product Development" course online except for the final examination.

The "Product Engineering" course was a full-credit course offered by HKU to the first year undergraduate students mainly for the "Industrial and Manufacturing Systems Engineering" programme. The course was evolved from a half-credit "Design for Manufacture" course several years ago. The course syllabus has also been updated several times and recently, it has been modified again to form a new course "Product Development". However, the basic aim of the course remains more or less the same. That is to motivate the students to proactively develop their integrative skills for product development. TELD is extensively used to deliver this new "Product Development" course. Significant activities include:

- Use TELD to Prepare and Host Course Materials
- Use TELD as an Virtual Classroom to Deliver Curriculum Materials Electronically
- Use TELD as an Virtual Study Room to Conduct Revisions and Assessments

The above three aspects are discussed very briefly in this section because of limitation in space. More detailed discussions are reported separately (Huang, Shen and Mak, 2001a, b, and 2002).

14.3.1.   Use *TELD* to Prepare and Host Course Materials

TELD provides a set of online facilities for teaching staff to prepare a variety of course materials. In the case of the "Product Engineering" and "Product Development" courses, the following activities have been conducted:

- The "Product Engineering" course has been created in TELD as a public course in the course library. The course, with some modifications in the syllabus has now been adapted within the TELD system online to form a new course, called "Product Development". Such course adoption involves the adoption of the course syllabus, curriculum and schedule. Further changes can be made at later stages with appropriate editors.
- The course curriculum of the new "Product Development" course is a duplication of the "Product Engineering" course curriculum right after adoption. Some examples are retained in the case study while others are either modified or newly created using the example editor as shown in Figure 14.5. Similarly, exercises in the mini-project are being updated according to the specific requirements of the new course. Case Editing is probably the most complicated operation in using TELD. Further discussions are given separately (Huang, Shen and Mak, 2001a).
- The schedule for the new "Product Development" course is changed because it is now delivered in the first semester while the "Product Engineering" course was delivered throughout the two semesters. Therefore, their schedules are very different.

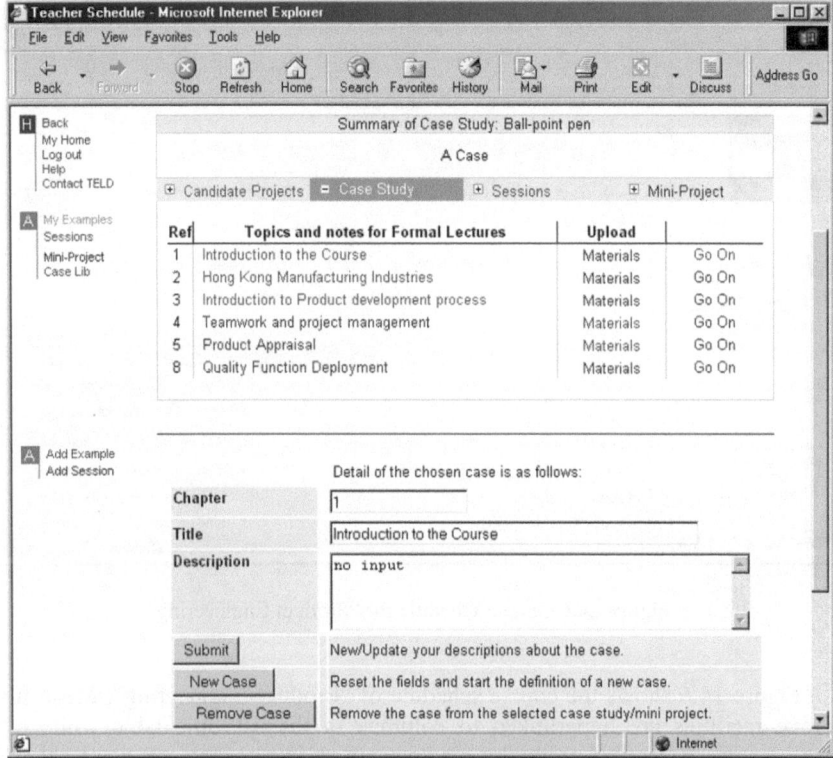

Figure 14.5 Example editor for creating and uploading teaching materials to TELD.

14.3.2. Use *TELD* as Virtual Classroom to Deliver Curriculum Materials Electronically

The TELD method and system has been pre-dominantly based on our experience in delivering this course over the last few years. TELD has now been used for delivering this course online except for the final examination. (Repeated) TELD provides a number of online facilities for delivering lectures and conducting activities associated with course delivery. Some of them are listed below and will be examined in more detail in this section:

- Checking participation and attendance in classes and workshops.
- Delivering lectures using the course materials according to the schedule.
- Conducting in-class discussions.
- Marking discussion questions (homework).
- Marking mini-project exercises.
- Conducting peer-assessment among student groups.
- Obtaining student feedbacks through the course forum and questionnaires.

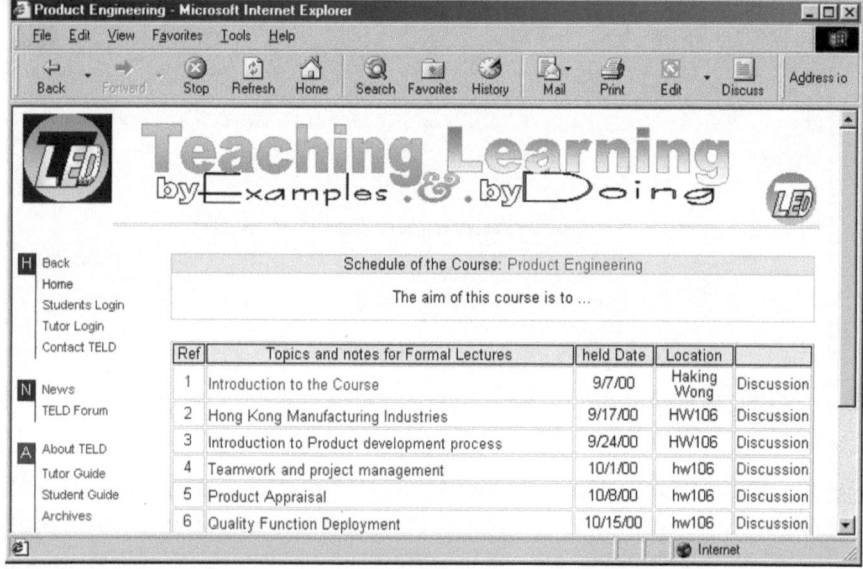

Figure 14.6 Course schedule for "Product Engineering".

Figure 14.6 shows the course schedule of "Product Engineering". Most of the above facilities are hyperlinked to columns in the schedule table while other facilities are accessible from the menu items shown on the left-hand side. For example, a click on the topic title will download the lecture notes, which are usually Powerpoint slide shows. A click on the lesson/session number will allow the tutor to take register in that class. A click on "In-Class Discussion" will show a list of discussion questions. Further discussion is given separately (Huang, Shen and Mak, 2001b).

14.3.3. Use *TELD* as Virtual Study Room to Conduct Revisions / Assessments

TELD plays a role of virtual study room by providing a set of online facilities for student-centred and self-learning activities. Further discussion is given separately (Huang, Shen and Mak, 2002). For example, after revising the examples discussed in the lessons, students should be able to answer the discussion questions set out by the lecturer. This is very straightforward. A student can get a list of discussion questions associated with the topic by clicking the "Assessments" hyperlink, next to the topic (example). To answer a question, just click over its description and then type the answer into the appropriate textbox. After submission, the tutor will mark and comment on it.

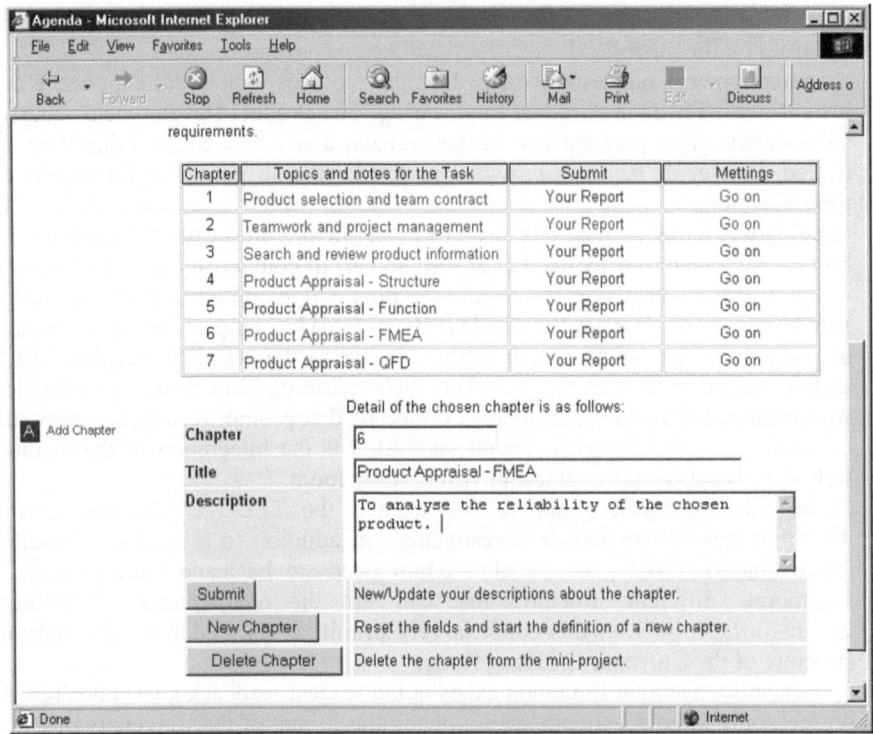

Figure 14.7 Online facilities for preparing mini-project plan.

TELD provides more sophisticated facilities for supporting the mini-projects. A group may use TELD to produce an overall project plan and prepare the agenda of a group meeting. After each meeting, TELD can be used to prepare the minutes of the meeting recorded as the progress report by the group. The project group may even upload the project report onto the TELD, chapter by chapter, for the tutor to assess the project work online. Figure 14.7 shows the sample screen for editing the agenda of a group meeting.

14.4.   SUMMARY

This chapter has presented the online TELD courseware engine over the World Wide Web. TELD refines the essence of various teaching and learning methods, such as the Case Method, the Project-based and Problem-based methods. TELD is a reflection of an ancient Confucius philosophy "I hear and I forget. I see and I remember. I do and I understand" in modern education within a technology-intensive environment. As an online courseware search engine and host, the TELD system incorporates the TELD method on the one hand, and provides a variety of facilities and tools for both the students and teachers in their teaching and learning activities on the other hand.

Although it is not tested widely for other courses, our initial experience and insights with TELD in "Product Engineering" and Product Development" courses have shown great potentials to be generalized and extended for other courses. Indeed, the set of facilities provided in TELD has no particular limitations on generalization. In comparison with the NEEDS, which is a search engine for teaching and learning materials, the TELD engine provides extra functionality for hosting such materials by the TELD web server. In comparison with the WebCT, which is also a web server for hosting teaching and learning materials, the TELD search engine incorporates the TELD (PBL or CM) teaching and learning method, in addition to the online search engine similar to the NEEDS. Besides, TELD assists tutors and students to plan their teaching and learning activities. Assessments by tutors are conducted online so that prompt feedback is provided between tutors and students. Therefore, TELD is the integration of the method, web server host, search engine and virtual study room.

In order to further improve and expand the TELD capabilities, several directions deserve immediate investigation, in addition to a number of routine improvements. Firstly, the Syllabus editor needs to be made more flexible to incorporate different structures that can suit the requirements of different universities. At present, this editor is very primitive, only adopting the syllabus structure of the University of Hong Kong.

Secondly, a similar limitation exists in the student feedback questionnaires. At present, only three questionnaires following the format of the University of Hong Kong are being used to obtain student feedback on the "Product Engineering" course, and lecturers and tutors (demonstrators) who are involved in the delivery of the course. A questionnaire editor would be necessary, so those different questions can be incorporated into the questionnaires to suit the requirements of other courses and universities.

Thirdly, lecturers' teaching notes for examples and students' reports on exercises are uploaded in their original formats (currently PDF, MS Word and Powerpoint files although other formats can also be supported). This proves sufficient for many teachers and students. However, this may not be adequate general for some other users. Also, it is not convenient for the TELD search engine to use the contents of these original files to find relevant course and case materials from the TELD database. It has been investigated to support the HTML format. However, it is not convenient to use the ordinary HTML files or pages for lecturers or tutors to give presentations in classes. Therefore, it is necessary to investigate a compromise.

Fourthly, the existing searching capabilities are limited. The existing basic search engine using keywords and categories is able to generate a list of relevant courses and cases. The user then "browses and picks" the course/case suitable for his/her application. This approach is sufficiently useful with limited size of the course/case libraries. When TELD is used for a wide variety of courses and a large number of cases, advanced search facilities are needed to filter the most relevant materials.

Fifthly, online examinations are not included in TELD. Some thoughts have been given to provide such facilities to support online exams. For example, examination sessions are arranged within a specified time period. The corresponding web pages containing the examination questions are made available to students who take part in the examination. However, there are technical and social constraints of doing this. Technically, students often need to answer questions using diagrams and equations. Students with limited skills of using computerised sketching and equation editing facilities may be unnecessarily disadvantaged. This may be compensated by giving them extra time. Socially, online examination may not be acceptable to many universities because of difficulties in preventing plagiarism. This is particularly afflicting where examinations take place at distance without close invigilation, even with examinations of the open-book type. One possible solution is to use video conferencing facilities to monitor the student identities and have online communications with the invigilators. Alternatively, students are required to attend examinations at specified exam centres where strict invigilation is maintained.

Sixthly, although the TELD engine is not yet intended as a commercial system, it does have the potential to fulfil this role to serve wider education communities. Therefore, an appropriate business model is required for TELD to obtain sufficient income to support the maintenance and improvement of the TELD system by itself. One model is in the form of consortium of corporate users (e.g. universities and education institutions). In this model, only the number of courses is controlled while the number of individual tutor and student users is not limited. For example, a diamond corporate membership allows a maximum of, say, 1,000 individual courses. A gold corporate membership accommodates 500 individual courses. A silver corporate membership supports 100 individual courses. In another business model, individual tutor users are registered. The number of courses that each registered tutor can create is controlled by the type of membership while the number of student users is not limited. For example, a diamond individual membership supports 10 courses. A gold individual membership supports 5 courses. A silver individual membership supports 2 courses. By using these two business models, different rates of subscription are then charged to different types of membership.

Seventhly, peer evaluation of courseware entered into TELD is not incorporated. This can be achieved in two ways. One is to require the courseware developers to provide self-evaluation of their materials by filling in a pro-forma form. The other is to invite third-parties to provide peer opinions on or experience with the courseware. Once again, a set of criteria may be organized into a pro-forma, as used in the NEEDS system. Such evaluation information would be invaluable for new users to make decisions if a course/case should be adopted/adapted for a particular course.

Finally, TELD is meant to support PBL and Case Method. The current case (example and exercise) editor achieves this objective only to a limited extent. Much remains for the tutors who actually prepare the course curriculum.

Technically, the above issues can be addressed by incorporating appropriate facilities into TELD. In fact, on-going efforts are being made in these areas. It is expected that they would be provided in the next release in the near future. TELD with such enhanced facilities is expected to play increasingly important roles in the future education within the so-called "Virtual or Digital University" environment.

LIST OF REFERENCES

Aberdeen (2000) Beating the competition with collaborative product commerce: Leveraging the Internet for new product innovation, Aberdeen Group, Inc., One Boston Place, Boston, Massachusetts, USA.

ACORN: Advanced Collaborative Open Resource Network, http://www.edrc.cmu.edu/ thrusts/flyers/acorn.html

Adiga, S. (1993) Object-oriented software for manufacturing systems, London: Chapman and Hall.

Afuah, A and Tucci, CL (2000). *Internet Business Models and Strategies: Text and Cases*, McGraw-Hill Higher Education. http://www.mhhe.com/business/ management/.

Ames, A.L., Nadeau, D.R., Moreland, J.L. (1996), *VRML 2.0 SourceBook*, Wiley, John & Sons, Incorporated.

Andreasen, M.M. (1987) "Design strategy", *Proceedings of ICED 87*, Boston, MA, ASME, 171-178.

Anton, A.I., Liang, E. (1996) "A Web-based requirement analysis tool", In: Proceedings of WET ICE'96, 238-243

Aranjo, C.S., Benedetto, H., Campello, A.C., Segre, F.M., Wright, I.C. (1996) "The utilisation of product development methods: A survey of UK industry", *Journal of Engineering Design*, Vol 7, No 3, 265-277.

Bacon, G., Beckman, S., Mowery, D.C., Wilson, E (1994) "Managing Product Definition in High-Technology Industries: A Pilot Study," *California Management Review*, 36/3 (Spring 1994): 32-56.

Bailey, MJ (1995) "Tele-Manufacturing: Rapid Prototyping on the Internet", *IEEE Computer Graphics and Applications*, November 1995, 20-26.

Balcerak, K.J., Dale, B.G. (1992) "Engineering Change Administration: the Key Issues", *Computer-Integrated Manufacturing Systems*, Vol. 5, No. 2, 125-132.

Barnes, L.B., Chistensen, C.R., Hansen, A.J. (1994) *Teaching and the Case Method*, published by Harvard Business School Press, Third Edition.

Barua, A and Whinston, A.B. (1999) "Measuring the Internet economy: An exploratory paper." *Working paper*, University of Texas, Austin, July, 1999, http://cism.bus.utexas.edu/works/articles/internet-economy.pdf

Beamon, B.M. (1998) "Supply chain design and analysis: Models and methods", *Inernational Journal of Production Economics*, Vol. 55, 281-294.

Bentley, R., Appelt, W., Busbach, U., Hinrichs, E., Kerr, D., Sikkel, K., Trevor, J., Woetzel, G. (1997) "Basic support for cooperative work on the World Wide Web", *International Journal of Human-computer Studies*, Vol 46, No. 6, 827-846.

Bentley, R., Applet, W. (1997) "Design a System for Cooperative Work on the World-Wide Web: Experiences with the BSCW System", *Proceedings of the 30th Hawaii International Conference on System Sciences, Maui, Hawii*, 7-10 January, 297-306

Berry, M.W., Browne, M. (1999) Understanding Search Engines, Mathematical Modeling and Text Retrieval, Siam.

Bidault, F., Despres, C., Butler, C. (1998) "New product development and early supplier involvement (ESI): the drivers of ESI adoption", *International Journal of Technology Management*, Vol.15, No.1/2, 49-69.

Bonaccorsi and Lipparini (1994) "Strategic partnerships in new product development: an Italian case Study", *Journal of Product Innovation Management*, Vol. 11, 134-145.

Boothroyd, G. (1996) "Book review on Design for Excellence by James G. Bralla", *Journal of Manufacturing Systems*, Vol. 15, No. 6, 443.

Boothroyd, G., Dewhurst, P., Knight, W. (1994) *Product design for manufacture and assembly*. Marcel Dekker Inc.

Boznak, R.G. (1993) *Competitive Product Development*, Milwaukee, WI: Business One Irwin/Quality Press.

Bryant, R.V.E., et al (1996) "Common product/process models for interfacing manufacturing simulation, process planning and CAD", In: *Proceedings of the 1996 ASME Conference*, Irvine, http://www.deneb.com/MADE/papers/dfm/dfm_96.html

BS 5760, Part 5, Guide to failure modes, effects and criticality analysis (FMEA and FMECA).

Burt, D (1989) "Managing suppliers up to speed", *Harvard Business Review*, Vol. 67, No. 4, july/August 1989, 127-135.

Bytheway, C.W. (1971) "The creative aspects of FAST diagramming", *Proceedings of the SAVE conference.*

Carter, D.E., Baker, B.S. (1992) Concurrent Engineering: The product development environment for the 1990s, Addison-Wesley Publishing Company.

Chakrabarti, A., Bligh, T.P. (1996), An approach to functional synthesis of mechanical design concepts: Theory, applications, and emerging research issues, *Artificial Intelligence for Engineering Design, Analysis and Manufacturing*, Vol. 10, 313-331.

Chang, E (1987) "Participant systems for cooperative work", *Distributed Arctifical Intelligence*, Huhns, MN (editor), Morgan Kanfmann Publishers, 311-339.

Choi, T.Y., Hartley, J.L. (1996) "An exploration of supplier selection practices across the supply chain", *Journal of Operations Management*, Vol.14, No., 333-343.

Clark, K (1989) "Project scope and project performance: The effect of parts strategy and supplier involvement on product development", *Management Science*, Vol. 35, No. 10, October 1989, 1247-1263.

Computer (1993) Special issue on Computer Supported Concurrent Engineering, Computer, January.

Cross, N. (1994) *Engineering Design Methods*, 2nd Ed. (Wiley).

Culley, S.J., Boston, O.P., McMahon, C.A. (1999) "Suppliers in new product development: Their information and integration", *Journal of Engineering Design*, Vol. 10, No. 1, 59-75.

Cutskosky, M.R., Tenenbaum, J.M., Glicksman, J. (1996) "MADEFAST: Collaborative engineering over the Internet", *Communications of the ACM*, Vol.39, No.9, 78~87.

Dahan, E., Srinivasan, V.S. (2000) "Reducing Market Risk for New Consumer Durables through Visual Depictions of Product Concepts," *Journal of Product Innovation Management,* Vol. 17, No. 2, 99-109.

Dale, B.G. (1982) "The Management of Engineering Change Procedure", *Engineering Management International*, Vol. 1, 201-208.

Dale, B.G., Shaw, P. (1990) "Failure mode and effect analysis in the motor industry: A state-of-the-art study", *Quality and Reliability Engineering International*, Vol. 6, No. 3, 179-188.

Decker, K.S. (1987) "Distributed problem-solving techniques: A survey", *IEEE Transactions on Systems, Man, and Cybernetics*, Vol 17, No 5, 729-740.

Dhillon, B.S., 1996, *Engineering Design: A Modern Approach*, Richard D. Irwin, Time Mirror Higher Education Group, Inc. Company.

Diprima, M. (1982) "Engineering Change Control and Implementation Considerations", *Production and Inventory Management Journal*, Vol. 23, Part 1, 81-87.

Dourish, P., Bellotti, V. (1992), Awareness and Coordination in Shared Workspaces, *CSCW 92 Proceedings*, 107-114.

Dowlatshahi, S. (1997) "The role of product design in designer-buyer-supplier interface", *Production planning & control*, Vol.8, No.6, 522-532.

Englemore, R, Morgan, T (eds) (1988) *Blackboard Systems*, Addison-Wesley Publishing.

Erkes, J.W., Kenny, K.B., Lewis, J.W. (1996), "Implementing Shared Manufacturing Services on the World Wide Web", *Communications ACM*, Vol. 39, No. 2, 34-45.

Erskine, J.A. Leenders, M.R., Mauffette-Leenders, L.A. (1998) *Teaching With Cases,* School of Business Administration, The University of Western Ontario.

Faulkner, P.D. (1997) "Statistical process control of test results via the Internet", In: *Proceedings of National Electronic Packaging & Production Conference*, 1997, 1159-1165.

Fedewa, C.S. (1996) "Business models for Internetpreneurs", 1996, http://www.gen.com/ iess/articles/art4.html.

Fine, C.H., Whitney, DE (1996) "Is the make-buy decision process a core competence?" *Working Paper*, MIT Centre for Technology, Policy and Industrial Development, http://imvp.mit.edu/imvpfree/Fine/Make_buy.pdf.

Fleischer, Liker, J.K. (1998) "Supply Chain Involvement", Chapter 5, *Concurrent Engineering Effectiveness: Integrating Product Development Across Organisation*, Hanser Gardner Publications, 113-139.

Ford Motor Company, *Potential failure mode and effect analysis*, Instruction Manual, Form 1696, 1998.

French, M.J. (1985) *Conceptual Design for Engineers* (London, Design Council).

Frivold T.J, Lang R.F., Fong M.W. (1995) Extending WWW for synchronous collaboration, *Computer Networks and ISDN Systems*, Vol. 28, 69-75.

Frost, R., Cutkoski, M. (1996) "An agent-based approach to making rapid prototyping processes manifest to designers", *Proceedings of ASME Symposium on Virtual Design and Manufacturing*, Irvine, CA, USA, August, 1996.

Gellersen, HW, Gaedke, M. (1999) *IEEE Internet Computing*, Vol. 31, Jan.-Feb., 60-68.

Ghodsypour, S.H., O'Brien, C. (1998) "A decision support system for supplier selection using an integrated analytic hierarchy process and linear programming", *International Journal of Production Economics*, Vol.56-57, 1998, 199-212.

Gill, H. (1990) "Adoption of design science – why so slow?" *Journal of Engineering Design*, Vol 1, No 4, 289-295.

Gill, H. (1994) "From contract supplier to market leader: A case study", *Journal of Engineering Design*, Vol 5, No 3, 187-194.

Gould, C. (1998) Searching Smart on the World Wide Web, Tools and Techniques for Getting Quality Results, Library Solutions Press, Berkeley, California.

Hague, M.J., Bendiab, A. (1998), Tool for the Management of Concurrent Conceptual Engineering Design, *Concurrent Engineering: Research and Applications*, Vol. 6, No. 2, 111-129.

Hall R. W., Mathur A., Jahanian F., Prakash A., Rassmussen C. (1996) Corona: A Communication Service for Scalable, Reliable Group Collaboration Systems, *Computer Supported Cooperative Work' 96*, November, 140-149.

Hanneghan, M., Merabti, M., Colquhoun, G. (1996) The World-Wide Web as a Platform for Supporting Interactive Concurrent Engineering, *Proceedings of Advanced Information Systems Engineering, 8th International Conference, CaiSE 96, Heraklion, Crete, Greece*, pp301-318.

Haque, B.U., Belecheanu, R.A., Barson, R.J., Pawar, K.S. (1999) Towards the Application of Case Based Reasoning to Decision-Making in Concurrent Product Development (Concurrent Engineering).

Hardwick, M. and D.L. Spooner, 1998, STEP services for sharing product models in virtual enterprises, in: *Proceedings of DETC98, ASME Design Engineering Technical Conference*, Atlanta, USA, on CD-ROM, Ref. CIE5518.

Harhalakis, G. (1986) "Engineering Changes for Made-to-order Products: How an MRP II System Should Handle Them", *Engineering Management International*, Vol. 4, 19-36.

Harris, S.B, (1996) "Business Strategy and the Role of Engineering Product Data Management: a Literature Review and Summary of the Emerging Research Questions", *Part B: Journal of Engineering Manufacture, Proceedings of IMechE*, Vol. 210, 207-219.

Hegde, G.G., Kekre, S.H., Su, H. (1992) "Engineering Changes and Time Delays: A Field Investigation", *International Journal of Production Economics*, Vol. 28, 341-352.

Helper, S.R., Sako, M. (1995) "Supplier relationships in Japan and the United States: Are they converging?" *Sloan Management Review*, Spring 1995, 77-84.

Hiltz, S.R., Benbunan-Fich, R. (1997) "Supporting collaborative learning in asynchronous learning networks", Invited keynote address for UNESCO/Open University *Symposium on Virtual Learning Environments and the Role of the Teacher*, Milton Keynes, England, April 1997.

HKTAIGA (1996) http://hktaiga.ust.hk/main.htm

Holt, G.D. (1998) "Which contractor selection methodology?" *International Journal of Project Management*, Vol.16, No.3, 153-164.

Huang, GQ., Feng, X.B., Mak, K.L. (2001) "POPIM: Pragmatic Online Project Information Management for Collaborative Product Development", In: ., July 2001, Canada.

Huang, G.Q., Mak, K.L. (1997a) "Engineering Changes: A survey within UK manufacturing industries", In: *Proceedings of International Conference on Managing Enterprises*, Loughborough University, July 1997.

Huang, G.Q., Mak, K.L. (1997b) "Computer aids for engineering change management", In: *Proceedings of International Conference on Computer Aided Production Engineering*, Poland, June 1997.

Huang, G.Q. (1996) (Editor) *Design for X - Concurrent Engineering Imperatives*, Chapman & Hall, London.

Huang, GQ, Jiang, ZH (2002) "Fuzzy set theoretical approach to design review of fuel pumps", *Proceedings of IMechE Vol. 216 Part B: Journal of Engineering Manufacture*, 287-292.

Huang, G.Q., Huang, J, Mak, K.L. (2000a) "Agent-based workflow management in collaborative product development on the Internet", *International Journal of Computer Aided Design*, Vol. 32, No. 2, 133-144.

Huang, G.Q., Huang, J, Mak, K.L. (2000b) "Early Supplier Involvement in New Product Development on the Internet: Implementation Perspectives", *International Journal of Concurrent Engineering: Research and Application*. Vol. 8, No. 1, 40-49.

Huang, G.Q., Lo, V., Yee, W.Y., Mak, K.L. (2000) "A Methodology for Engineering Change Impact Analysis", IMechE Conference Transactions for the *16th International Conference on Computer Aided Production Engineering*, 7-9 August 2000, The University of Edinburgh, UK, 603-611.

Huang, G.Q., Mak, K.L. (1997a) "Developing a generic DFX shell", *Journal of Engineering Design*, Vol. 8, No. 3, 1997, 251-260.

Huang, G.Q., Mak, K.L. (1997b) "DFX: A generic shell for developing Design for X tools", *International Journal of Robotics and Computer Integrated Manufacturing*, Vol. 13, No. 3, 271-280.

Huang, G.Q., Mak, K.L. (1998a) "DFX: A generic shell for applying Design for X tools", *International Journal of Computer Integrated Manufacturing*, Vol. 11, No. 6, 475-484.

Huang, G.Q., Mak, K.L. (1998b) "Web based collaborative product development", Invited lecture, *1st International Seminar and Workshop on*

Engineering Design in Integrated Product Development, 8-10 October, 1998, Poland.

Huang, G.Q., Mak, K.L. (1999a) "Current Practices of Managing Engineering Changes in UK Manufacturing Industries", *International Journal of Operations and Production Management*, Vol. 19, No. 1, 21-37.

Huang, G.Q., Mak, K.L. (1999b) "Web-Based Morphological Chart Analysis for Innovative Conceptual Design in Collaborative Product Development", *International Journal of Intelligent Manufacturing*, Vol. 10, No. ¾, 267-278.

Huang, G.Q., Mak, K.L. (1999c) "Web-based quality function deployment", In: *Proceedings of International Conference on Computers and Industrial Engineering*, Melbourne, Australia, December 19999. 711-716.

Huang, G.Q., Mak, K.L. (1999d) "Web-Based Design for Manufacture and Assembly", *Computers in Industry: An International Journal*, Vol. 38, No. 1, 17-30.

Huang, G.Q., Mak, K.L. (2000a) "Modelling the Customer-Supplier Interface over the World Wide Web to Facilitate Early Supplier Involvement in New Product Development", *Institution of Mechanical Engineers Proceedings, Part B Journal of Engineering Manufacture*, Vol. 214, No. 9, 759-769.

Huang, G.Q., Mak, K.L. (2000b) "WeBid: A Web-Based Framework to Support Early Supplier Involvement in New Product Development", *International Journal of Robotics and Computer Integrated Manufacture*, Vol. 16, No. 2-3, 169-179.

Huang, G.Q., Mak, K.L. (2001a) "Issues in the Development and Implementation of Web Applications for Product Design and Manufacture", *International Journal of Computer Integrated Manufacture*. Vol. 14, No. 2, 125-135.

Huang, G.Q., Mak, K.L. (2001b) "Web-integrated manufacturing: Recent developments and Emerging issues", *International Journal of Computer Integrated Manufacture,* Vol. 14, No. 2, 3-13.

Huang, G.Q., Mak, K.L. (2001c) "Web-Based Electronic Product Cataloguing on the Internet", *International Journal of Computer Applications in Technology,* Vol. 14, Nos. 1/2/3, 17-39.

Huang, G.Q., Nie, M., Mak, K.L. (1999) "Web-based failure mode and effect analysis", *Computers and Industrial Engineering: An International Journal*, Vol. 37, No. 1-2, 177-180.

Huang, G.Q., Shen, B. Mak, K.L. (2001a) "TELD Courseware Engine as a Virtual Classroom for Active and Collaborative Teaching", *International Journal of Engineering Education* (for Special Issue on Virtual Universities). Vol 17, No. 2, 164-175.

Huang, G.Q., Shen, B. Mak, K.L. (2001b) "WWW.TELD.Net: Online Courseware Engine for Teaching by Examples and Learning by Doing", *Journal of Educational Technology Systems,* Vol. 29, No. 3, 219-235.

Huang, G.Q., Shen, B. Mak, K.L. (2002) "Participatory and Collaborative Learning with TELD Courseware Engine", *ASCE Journal of Professional Issues in Engineering Education and Practice*, Vol. 128, No. 1, 36-43.

Huang, G.Q., Shen, B., Mak, K.L. (2000) "WAPIP: Web Applications in Product Introduction Process", *International Journal of Advanced Manufacturing Technology*, Vol. 17, No. 10, 775-782.

Huang, G.Q., Shi, J, Mak, K.L. (2000a) "Web-based design for x guidelines", *International Journal of Material Processing Technology*, Vol. 107, 71-78.

Huang, G.Q., Shi, J, Mak, K.L. (2000b) "Failure mode and effect analysis over the world wide web", *International Journal of Advanced Manufacturing Technology*, Vol. 16, No. 8, 603-608.

Huang, G.Q., Yee, W.Y., Mak, K.L. (2000) "Re-Engineering the Engineering Change Management Process", In: *Proceedings of the 7ᵗʰ International Conference on Concurrent Engineering: Research and Applications*, Lyon Claude Bernard University, France, July 17-20 2000, 255-264.

Huang, G.Q., Yee, W.Y., Mak, K.L. (2001) "Development of a web-based system for engineering change management", *International Journal of Robotics and Computer Integrated Manufacturing*, Vol 17, No. 3, 255-267.

Hubka, V., Andreasen, M.M., Eder, W.E. (1998) *Practical Studies in Systematic Design*, Butterworths, London.

IAMS, Intelligent Assembly Modelling and Simulation: IAMS, http://www.ndim.edrc.cmu.edu/CODES/Project/intelligent.html.

Ichida, T., 1989, *Deizain Rebyu Jireishu*, Productivity Press, Portland, Oregon.

IP3S, Integrated Process Planning / Production Scheduling, http://agile.cimds.ri.cmu.edu/IP3S_home.html.

IPPI, Integrated Product Processing Initiative, http://agile.cimds.ri.cmu.edu/IPPI /IPPI_executive.html.

Jones, V.C. (1987) MAP/TOP networking – A foundation for computer integrated manufacturing, McGraw-Hill Book Company.

Kalyanapasupathy, V., Lin, E., Minis, I. (1997) "Group Technology code generation over the Internet", http://www.isr.umd.edu/Labs/CIM/profiles/lin/ docs/gt/.

Karlsson, C.; Nellore, R.; Soderquist, K. (1998) "Black box engineering: Redefining the role of product", *Journal of Product Innovation Management*, Vol. 15, No. 6, 534-549.

Kim, C.Y., Kim, N., Kim, Y., Kang, S.H., O'Grady, P. (1998) "Distributed Concurrent Engineering: Internet-Based Interactive 3-D Dynamic Browsing and Markup of STEP Data", *Concurrent Engineering: Research and Applications*, Vol. 6, No. 1, 53-70.

Kim, N., Kim, C.Y., Kim, Y., Kang, S.H., O'Grady, P. (1996) "Collaborative Design using the world wide web", http://www.iil.ecn.uiowa.edu/Internetlab/ Techrep/HTML/, TR9702.htm.

Klein, M (1992) "Detecting and resolving conflicts among cooperating human and machine-based design agents", *International Journal of Artificial Intelligence in Engineering*, Vol 7, No 4, 1992, 93-104.

Klein, M. (1995), Core Services for Coordination in Concurrent Engineering, Proceedings of the Fourth Workshop on Enabling Technologies: Infrastructure for Collaborative Enterprises, 1995, 189-198.

Klein, M., Lu, S.C-Y., Baskin, A.B. (1990), Towards a Theory of Conflict Resolution in Cooperative Design, *Proceedings of the Twenty-Third Annual Hawaii International Conference on System Sciences, 1990*, Vol. 4, 41-50.

Kroemker, M., Thoben, K.-D., Wickner, A. (1997) "An infrastructure to support concurrent engineering in bid preparation", *Computers in Industry*, Vol. 33, 201-208.

Laliberty, Thomas J. Hildum, David W., Sadeh, Norman M., McA'Nulty, John, Kjenstad, Dag Bryant, Robert V.E. and Smith, Stephen F. (1996) "A Blackboard Architecture for Integrated Process Planning/Production Scheduling", *Proceedings of ASME Design for Manufacturing*. August, 1996.

Lamming R., et al (1999) "Project ION - Literature review of supply networks", http://www.labs.bt.com/people/callagjg/ion/supply.htm.

Lee, J. and Lai, K.Y. (1991) "What's In Design Rationale?" *Human-Computer Interaction*, Vol. 6, No. 3-4, 251-280.

Lee, S.W., Huang, G.Q., Mak, K.L. (1999) "A synchronous Internet-based system for requirement analysis in collaborative product definition", In: CD Proceedings of 4th International Conference on Industrial Engineering Applications and Practice, November 1999. Paper ID: 117SL&GH&KM.

Lee, W.B. (2000) "Digital Factory: Manufacturing in Information Era", *China Mechanical Engineering*, Vol. 11, No. 1-2, 93-96 (In Chinese).

Leech, D.J., Turner, B.T. (1985) *Engineering Design for Profit*, Ellis Horwood Limited. Chichester, England (Chapter 12).

Mahadevan, B (2000) "Models for Internet based e-commerce: An anatomy", California Management Review, Vol. 42, No. 4., 55-69.

Manufacturing Foresight 2020, http://www.dti.gov.uk/manufacturing/foresight.pdf

Maull, R., Hughes, D., Bennett, J. (1992) "The Role of the Bill-of-Materials as a CAD/CAPM Interface and the Key Importance of Engineering Change Control", *Computing & Control Engineering Journal*, March 1992, 63-70.

McClure, C. (1992) The Three Rs of Software Automatio: Re-engineering, Repository, Reusability, Prentice Hall.

McQuater, R.E., Dale, B.G., Boaden, R.J., Wilcox, M. (1996) "The effectiveness of quality management tools and techniques: An examination of the key influences in five plants", *IMechE Proceedings, Part B Engineering Manufacture*, Vol. 210, No. B4, 329-339.

Miles, L.D. (1965) *Techniques of Value Analysis*, 2nd Ed. (McGraw Hill).

Miller, J.A., Palaniswami, D., Sheth, A.P., Kochut, K.J., Singh, H. (1997) "WebWork: METEOR2's Web-based workflow management system", *Journal of Intelligent Information Systems*, Vol. 10, 185-215.

Mills, A. (1998) *Collaborative Engineering and the Internet*, Society of Manufacturing Engineers.

MIL-STD1629A, Procedures for performing a failure mode, effects, and criticality analysis, US Department of Defense.

Minis, I., et al (1995) "Optimal selection of partners in agile manufacturing", http://www.isr.umd.edu/Labs/CIM/.

Monahan, R.E. (1995) Engineering Documentation Control Practices and Procedures, Marcel Dekker, Inc, New York, USA.

Muller, P.C., de Poorter, R., de Jong, J., van Engelen, J.M.L. (1996) "Using the Internet as a communication infrastructure for lead user involvement in the new product development process", In: *Proceedings of WET ICE '96*, 220-225

Nicholls, K. (1990) "Getting Engineering Changes Under Control", *Journal of Engineering Design*, Vol. 1, No 1, 1-6.

Nii, H.P. (1986) "Blackboard systems: The blackboard model of problem solving and the evolution of blackboard architectures", *AI Magazine*, Vol 7, No 2.

Norell, M. (1993) "The use of DFA, FMEA, and QFD as tools for concurrent engineering in product development processes", In: *Proceedings of ICED 93*, The Hague.

Norris, K.W. (1963) "The morphological approach to engineering design", In: J.C. Jones and D. Thornley (eds), *Conference on Design Methods*, Pergamon, Oxford.

Orfali, R, Harkey, D (1998) *Client/Server Programming with Java and Corba*, Second Edition, Wiley Computer Publishing, John Wiley & Sons, Inc.

Pahl, G. Beitz, W. (1984) *Engineering Design* (London, Design Council).

Pandey, A., Clausing, D.P. (1991) "QFD implementation survey report", *Working Paper, Laboratory for Manufacturing and Productivity*, MIT, Cambridge, November.

Parkinson, J (1999) "Retail models in the connected economy: Emerging business affinities", 1999 http://www.ey.com/global/gcr.nsf/us/insights_-_eBusiness_-_Ernst_&_Young_LLP.

Parunak, HVD (1988) "Manufacturing experience with the contract-net", *Distributed Arctifical Intelligence*, Huhns, MN (editor), Morgan Kanfmann Publishers, 285-310.

Petrie, C.J. (1996) "Agent-based engineering, the Web, and intelligence", *IEEE Expert*, December 1996, 24-29.

Pfund, A.P., 2001, Design on an International Project, In: *Proceedings of 2001 ASME DETC, International Issues in Engineering Design*, Pittsburgh, USA, September 2001.

Pham, D.T. (1998) http://intell-lab.engi.cf.ac.uk/manufacturing/ipm/ipm.html

Pugh, S. (1991) Total design: Integrated methods for successful product engineering (Addison-Wesley).

RaDEO (1997) http://radeo.nist.gov/radeo/.

Rampersad, H.K. (1994) Integrated and Simultaneous Design for Robotic Assembly (Wiley).

Reidelbach, M.A. (1991) "Engineering Change Management in Long-lead-time Environments", *Production and Inventory Management Journal*, Vol. 32, No. 2, 84-88.

Reinders, H. (1995) "Design information deployment in 'Design Assistants'", In: *Proceedings of ICED 95*, Praha, Czech Republic, 1339-1344.

Rezayat, m. (2000) "The Enterprise-Web portal for life-cycle support", *Computer-aided Design*, Vol. 32, No. 2, 85-96.

Romano, N.C., Nunamaker, J.F., Briggs, R.O., Vogel, D.R. (1998) "Architecture, design, and development of an HTML/JavaScript web-based group support system", *Journal of the American Society of Information Science*, Vol. 49, No. 7, 649-667.

Roozenburg, N.F.M., J. Eekels (1995) *Product design: fundamentals and methods*, Wiley series in product development, Chichester, West Sussex, England.

Roseman M., Greenberg S. (1996) TeamRooms: Network Places for Collaboration, *Computer Supported Cooperative Work' 96*, November, 325-333.

Roy, R., Potter, S. (1996) "Managing engineering design in complex supply chains", *International Journal of Technology Management*, Vol.12, No.4, 403-420.

Roy, U., Bharadwaj, B., Kodkani, S.S., Carian, M. (1997) "Product development in a collaborative design environment", *Concurrent Engineering: Research and Applications*, Vol. 5 No. 4, 347-365.

Roy, U., Kodkani, S.S. (2000) Collaborative Product Conceptualisation Tool using Web Technology, *Computers in Industry*, Vol. 41, Issue 2, 195-209, March 2000.

Saeed, I., Bowen, D.M., Sohoni, V.S. (1993) "Avoiding engineering changes through focused manufacturing knowledge", *IEEE Transactions on Engineering Management*, 40 (1), 54-58.

Schlachter, E (1995) "Generating revenues from websites", Board Watch, July 1995, http://boardwatch.internet.com/mag/95/jul/bwm39.html .

Schoonmaker, S.J., 1996, *ISO 9001 for Engineering and Designers*, The McGraw-Hill Companies, Inc.

Shi, J, Huang, G.Q., Mak, K.L. (1999) "MetaDFX: A web-based system for developing design for x tools" In: *CD Proceedings of 4th International Conference on Industrial Engineering Applications and Practice*, November 1999.Paper ID: 106JS&GH&KM.

Shi, J, Huang, G.Q., Mak, K.L. (2001) "Performance measurement in web-based design for x", *International Journal of Computers in Industry*. Vol 44, No. 1, 67-78.

Sikkel, K., Process Support for Cooperative Work on the World Wide Web, Proceedings of the Sixth Euromicro Workshop on Parallel and Distributed Processing, 1998, 325-331.

Smith, C.S., Wright, P.K. (1996) "CyberCut: A world wide web based design to fabrication tool", *Journal of Manufacturing Systems*, Vol. 15, No. 6, 432-442.

Sonnenwald D. H. (1996) Communication roles that support collaboration during the design process, *Design Studies*, Vol. 17, 277-301.

Spink, A., Bateman, J., Jansen, B.J. (1999) "Searching the Web: a survey of EXCITE users", http://www.emerald-library.com/brev/ 17209bel.htm

Sturges, R.H., O'Shaughnessy, K., Reed, R.G. (1993) "A systematic approach to conceptual design", *Concurrent Engineering: Research and Applications*, 1, 93-105.

Swank, M, Kittel, D, Spenik, D (1997) *Web Database Developer's Guide with Visual Basic 5,* Sams.net Publishing.

Swift, K.G. (1981) *Design for Assembly Handbook*, Salford University Industrial Centre Ltd., UK.

Sycara, K. (1989) "Multi-Agent Compromise via Negotiation". In *Distributed Artificial Intelligence* (Vol. 2), Gasser, L. and Huhns, M. (Eds.), Morgan Kaufman Publishers, Los Altos, CA., September, 1989.

Technicomp (1990) *Application Guide of Failure Mode and Effect Analysis*, Technical Competence Through Training.

Timmers, P (1998) "Business models for electronic markets", Electronic Markets, Vol. 8 No. 2, 3-8, 1998.

Timmers, P (2000) *Electronic Commerce: Strategies and Models for Business-To-Business Trading*, Wiley Series in Information Systems, John Wiley & Sons, Chichester, UK.

Tong, C (1987) "Towards an engineering science of knowledge based design", *International Journal of Artificial Intelligence in Engineering*, Vol 2, No 3, 1987, 133-166.

Tseng, M.M., Jiao, J. (1997) "A Variant Approach to Product Definition by Recognising Functional Requirement Patterns", *Computers in Engineering*, Vol. 33, Nos. 3-4, 629-633.

Tuikka, T., Salmela, M. (1998), Facilitating designer-customer communication in the World Wide Web, *Internet Research: Electronic Networking Applications and Policy*, Vol. 8, No. 5, 442-451.

Turoff, M. (1994) "The marketplace road to the information highway", http://www.nijt.edu/Virtual_Classroom/Papers/Market.html

Twigg, D. (1998) "Managing product development within a design chain", *International Journal of Operations & Production Management*, Vol.18, No.5, 508-524.

Ulrich, K.T., Eppinger, S.D. (1995) *Product Design and Development*, McGraw-Hill.

Vanwelkenhuysen, J. (1998) "The tender support system", *Knowledge-Based Systems*, Vol. 11, 363-372.

Verma, R., Pullman, M. E. (1998) "An analysis of the supplier selection process", *Omega, Int. Journal of Management Science*, Vol. 26, No. 6, 739-750.

VICS CPFR available at: www.cpfr.org.

VM Phase II, http://www.isr.umd.edu/Labs/CIM/projects/virtual.html

Voigt, E.C. (1996), Product design review, a method for error-free product development, Productivity Press, Portland, Oregon.

Vokurka, R.J., Choobineh, J. and Vadi, L. (1996) "A prototype expert system for the evaluation and selection of potential suppliers", *International Journal of Operations & Production Management*, Vol. 16, No. 12, 106-127.

Wade, V.P., Grimson, J.B., Power, C. (1999) "WWW-based educational environments for software engineers", *International Journal of Engineering Education*, Vol. 15, No. 2, 130-136.

Wagner, R., Castanotto, G., Goldberg, K. (1995), "DFX via the Internet", *SPIE*, Vol. 2596, 192-195. Also FixtureNet at http://teamster.usc.edu/fixture/

Watts, F. (1984) "Engineering Changes: A Case Study", *Production and Inventory Management Journal*, Vol. 25, Part 4, 55-62.

Whitney, D. et al (1995) "Agile pathfinders in the aircraft and automobile industries – A progress report", *Internal Working Report*, MIT.

Will, P. (1996) *Active Catalogs Project Home Page*, http://www.isi.edu/active-catalog/index.html

Wong, CC, Veeramani, D., Chalermdamrichai, V. (1996) "QUESTER: A Computer-Integrated System for Virtual Shopping Through the Internet", *5th IE Research Conference Proceedings*, Minneapolis, MN, pp. 758-763, May 18-19.

Wright, D.T., Burns, N.D. (1997) "Rapid prototyping cellular systems for virtual global enterprises", In: *Proceedings of 5th International Conference on FACTORY 2000*, 2-4 April 1997, IEEE Conference Publications No. 435, 423-428.

Wright, I.C. (1997) "A Review of Research into Engineering Change Management: Implications for Product Design", *Design Studies*, Vol. 18, No 1, 33-42.

Wright, I.C. (1998) Design Methods in Engineering and Product Design, McGraw Hill.

X-CITTIC, Virtual Enterprise Supply Chain Planning and Control

Yee, W.Y., Huang, G.Q., and Mak, K.L. (2000) "Current practice of engineering change management in Hong Kong manufacturing industries", *The ninth International Manufacturing Conference in China*, 16-17 August 2000, Hong Kong.

Yee, W.Y., Huang, G.Q., Mak, K.L. (1998) "Towards a Reference Framework for Engineering Change Management", *3rd International Conference on Industrial Engineering*, Hong Kong University of Science and Technology, Hong Kong, December, 1998.

Yen, B.P.C. (1997a) "An interactive scheduling agent on the Internet", In: *Proceedings of the IEEE 30th Annual Hawaii International Conference on System Science (HICSS-30)*, Hawaii, January 7-10, 1997.

Yen, B.P.C. (1997a) "Web-based simulation tools", In: Proceedings of 4th International Conference on Industrial Engineering and Engineering Management, August 18-22 1997, Hong Kong.

LIST OF WEBSITES

http://about.webct.com/library/v3_white.html
http://agile.cimds.ri.cmu.edu/
http://aims.parl.com/AIMSNet-Features.html
http://cism.bus.utexas.edu/works/articles/internet-economy.pdf
http://ecommerce.ncsu.edu/topics/models/models.html "Business models on the web"
http://edweb.sdsu.edu/clrit/learningtree/PBL/WhatisPBL.html
http://herkules.appsolut.com/eps/en/PDFs/Whitepaper A4.PDF
http://hktaiga.ust.hk/main.htm
http://intell-lab.engi.cf.ac.uk/manufacturing/ipm/ipm.html
http://java.sun.com/products/java-server/servlets/
http://java.sun.com/products/jdbc/index.html
http://java.sun.com/products/jdk/rmi/index.html
http://radeo.nist.gov/radeo/.
http://urobe.uni-paderborn.de/GEN/
http://www.aboutportals.com/technology/
http://www.appsolut.com/Default.htm/
http://www.autodesk.com/foundation/pbl/
http://www.blackboard.com
http://www.cpfr.org/
http://www.dti.gov.uk/manufacturing/foresight.pdf
http://www.ecch.cranfield.ac.uk/
http://www.ecollege.com
http://www.e-education.com
http://www.execpc.com/~gopalan/misc/compare.html
http://www.fmeca.com.
http://www.frufalfun.com/5webincomemodels.html
http://www.gen.com/iess/articles/art4.html
http://www.hbsp.harvard.edu/home.html
http://www.hiperworld.com/business-model.html
http://www.howarddowding.com/

http://www.iil.ecn.uiowa.edu/Internetlab/Techrep/HTML/
http://www.insurance.about.com/industry/insurrance/library/weekly/aa042000a.htm
http://www.isi.edu/active-catalog/index.html
http://www.isr.umd.edu/Labs/CIM/profiles/lin/docs/gt/.
http://www.isr.umd.edu/Labs/CIM/vm/
http://www.labs.bt.com/people/callagjg/ion/supply.htm
Http://www.learnasp.com/.
http://www.ndim.edrc.cmu.edu/CODES/Project/intelligent.html
http://www.needs.org/
http://www.nijt.edu/Virtual_Classroom/Papers/Market.html
http://www.nimblesite.com/xcittic/default.htm
http://www.portalwave.com/portal_wave/application.html
http://www.projectsonline.com/pol_content/pages/body_pds.htm
http://www.samford.edu/pbl/pbl_main.html
http://www.slate.com/
http://www.supply-chain.org/
http://www.webct.com
http://www.workz.com/content/1148.asp
http://www.cpfr.org
http://www.comp.lancs.ac.uk/edc/schemebuilder/

INDEX